ENERGY SCIENCE, ENGINEERING AND TECHNOLOGY

THE POTENTIAL FOR WOOD ENERGY AND BIOMASS FUELS IN ALASKA

ENERGY SCIENCE, ENGINEERING AND TECHNOLOGY

Additional books in this series can be found on Nova's website under the Series tab.

Additional E-books in this series can be found on Nova's website under the E-books tab.

ENERGY SCIENCE, ENGINEERING AND TECHNOLOGY

THE POTENTIAL FOR WOOD ENERGY AND BIOMASS FUELS IN ALASKA

DMITRY S. HALINEN
EDITOR

Nova Science Publishers, Inc.
New York

Library of Congress Cataloging-in-Publication Data

The potential for wood energy and biomass fuels in Alaska / editor, Dmitry S. Halinen.
    p. cm.
  Includes index.
  ISBN 978-1-61470-990-9 (hardcover)
  1. Biomass energy--Alaska. 2. Fuelwood--Alaska. I. Halinen, Dmitry S.
  TP339.P66 2011
  662'.88--dc23
                        2011030096

Published by Nova Science Publishers, Inc. ✝ *New York*

CONTENTS

PREFACE

Goal three of the current U.S. Department of Agriculture, Forest Service strategy for improving the use of woody biomass is to help develop and expand markets for woody biomass products. This book explores the existing volumes of renewable wood energy products (RWEP) that are currently used in Alaska and the potential demand of RWEP for residential and community heating projects in the state. Data published by the U.S. Department of Commerce, Bureau of Census and the U.S. Department of Energy have been used to build a profile of residential and commercial energy demand for Alaska census tracts. If fuel oil prices increase to the levels experienced in 2008, there would be a strong economic incentive to convert heating systems to use solid wood fuels

Chapter 1 - Rural Alaskan communities are faced with the concurrent problems of high fuel prices for electricity and heating, high fire risk owing to increasing fire severity and fuel buildup around communities, and environmental contamination from extensive use of diesel fuel. In this study author sought partial solutions for all of these problems through use of wood energy in rural Alaskan villages in forested regions of interior Alaska. Author assessed the feasibility of this fuel substitution from an ecological, economic, and social viewpoint using separate submodels, then analyzed the author's results as a whole.

Chapter 2 - The current U.S. Department of Agriculture, Forest Service strategy for improving the use of woody biomass defines four strategy goals. Goals one, two, and four include building partnerships, developing and deploying science and technology, and assuring a supply of biomass. Goal three of the strategy is to "help develop new and expanded markets for bioenergy and biobased products". All goals are viewed by the Forest Service as important parts of a primary and broader objective of sustaining healthy forests that will survive natural disturbances and threats, including climate change.

In: The Potential for Wood Energy ...   ISBN: 978-1-61470-990-9
Editor: Dmitry S. Halinen   © 2012 Nova Science Publishers, Inc.

Chapter 1

ASSESSING THE POTENTIAL FOR CONVERSION TO BIOMASS FUELS IN INTERIOR ALASKA[*]

Nancy Fresco and F. Stuart Chapin III

The Forest Service of the U.S. Department of Agriculture is dedicated to the principle of multiple use management of the Nation's forest resources for sustained yields of wood, water, forage, wildlife, and recreation. Through forestry research, cooperation with the States and private forest owners, and management of the National Forests and National Grasslands, it strives—as directed by Congress—to provide increasingly greater service to a growing Nation.

The U.S. Department of Agriculture (USDA) prohibits discrimination in all its programs and activities on the basis of race, color, national origin, age, disability, and where applicable, sex, marital status, familial status, parental status, religion, sexual orientation, genetic information, political beliefs, reprisal, or because all or part of an individual's income is derived from any public assistance program. (Not all prohibited bases apply to all programs.) Persons with disabilities who require alternative means for communication of program information (Braille, large print, audiotape, etc.) should contact USDA's TARGET Center at (202) 720-2600 (voice and TDD). To file a

---

[*] This is an edited, reformatted and augmented version of the United States Department of Agriculture publication, Forest Service, Pacific Northwest Research Station, Research Paper PNW-RP-579, dated June 2009.

complaint of discrimination, write USDA, Director, Office of Civil Rights, 1400 Independence Avenue, SW, Washington, DC 20250-9410 or call (800) 795-3272 (voice) or (202) 720-6382 (TDD). USDA is an equal opportunity provider and employer.

ABSTRACT

Fresco, Nancy and Chapin F. Stuart III. 2009. Assessing the potential for conversion to biomass fuels in interior Alaska. Res Pap. PNW-RP-579. Portland, OR: U.S. Department of Agriculture, Forest Service, Pacific Northwest Research Station. 56 p.

In rural Alaskan communities, high economic, social, and ecological costs are associated with fossil fuel use for power generation. Local concerns regarding fuel prices, environmental contamination, and the effects of global climate change have resulted in increased interest in renewable energy sources. In this study we assessed the feasibility of switching from fossil fuels to wood energy in rural Alaskan villages in forested regions of interior Alaska. Modeling results based on recent data on rural energy use, demographics, economics, and forest dynamics indicated that the installation costs of biomass systems would be recouped within 10 years for at least 21 communities in the region. In addition, results showed that all but the largest remote communities in the interior could meet all their electrical demand and some heating needs with a sustainable harvest of biomass within a radius of 10 km of the village. Marketable carbon credits may add an additional incentive for fuel conversion, particularly if U.S. prices eventually rise to match European levels. Biomass conversion also offers potential social benefits of providing local employment, retaining money locally, and reducing the risk of catastrophic wildfire near human habitation. This analysis demonstrated that conversion to biomass fuels is economically viable and socially beneficial for many villages across interior Alaska.

Keywords: Biomass fuel, carbon offset, interior Alaska, wood energy.

SUMMARY

Rural Alaskan communities are faced with the concurrent problems of high fuel prices for electricity and heating, high fire risk owing to increasing fire severity and fuel buildup around communities, and environmental contamination from extensive use of diesel fuel. In this study we sought partial solutions for all of these problems through use of wood energy in rural Alaskan villages in forested regions of interior Alaska. We assessed the feasibility of this fuel substitution from an ecological, economic, and social viewpoint using separate submodels, then analyzed our results as a whole.

Owing to the high costs of fuel transport and storage in rural Alaska, energy prices are extremely high. Consumers pay an effective rate of up to 35 cents per kWh in some regions, despite substantial state subsidies. Our focus was on use of black spruce (*Picea mariana* (Mill.) B.S.P.) in relatively simple small-scale boilers for electrical power generation, with the possibility of using waste heat in combined heat and power systems. In addition, we explored the possibility of communities obtaining carbon offset credits that could be traded on the open market.

The ecological submodel estimated the maximum travel distance necessary for biomass harvest for wood energy around each of the 36 villages studied. This submodel took into account village population, per capita energy use, the fraction of total energy use to be replaced with biomass energy, rotation length for forest harvest, biomass density for black spruce at harvest age, wood energy density, electrical efficiency, and percentage of forest cover.

The economic submodel explored the short- and long-term costs and benefits of switching from diesel energy to wood energy in these remote communities, and estimated the period needed to pay back capital investments. In our calculations, we used the installed cost of a biomass power system per kilowatt of generation capacity, the total biomass capacity installed, the actual energy offset, diesel efficiency, diesel price, the fraction of nonfuel costs offset by use of biomass, total nonfuel costs, biomass energy generated, biomass energy costs, and the value of carbon credits available owing to fuel offset.

We explored the effects of model input selection and model parameter uncertainty on model outputs by performing sensitivity analyses on both the ecological and the economic submodels.

Our social analysis was qualitative, and focused on factors likely to affect the feasibility of fuel substitution, including threshold requirements for success in any one community. We also examined potential feedback between ecological, economic, and social factors, and assessed ways in which they might in combination affect the feasibility of wood biomass fuel use in Alaska villages.

Our analysis was intentionally conservative, and may therefore have underestimated potential advantages of conversion to biomass fuels. Nevertheless, modeling results indicated that the installation costs of biomass systems would be recouped within 10 years for at least 21 communities in the region. In addition, results showed that all but the largest remote communities in the interior could meet all their electrical demand and some heating needs with a sustainable harvest of biomass within a radius of 10 km of the village. The greatest economic feasibility is demonstrated by villages that are not easily reached by either road or river networks. The greatest ecological feasibility occurs in communities of small to medium size, where the wood resources needed are available within a relatively small radius. Marketable carbon credits may add an additional incentive for fuel conversion, particularly if U.S. prices eventually rise to match European levels. Biomass conversion also offers potential social benefits of providing local employment, retaining money locally, and reducing the risk of catastrophic wildfire near human habitation. Success of a fuel conversion project in a community is likely to depend upon the existence of local advocates and participants; sufficient local technological skills; and collaboration among communities, funders, and electrical cooperatives.

This analysis demonstrated that conversion to biomass fuels would be economically viable and socially beneficial for many villages across interior Alaska. Pilot projects offer the next step in testing feasibility.

INTRODUCTION

The excess carbon dioxide released into the atmosphere by the burning of fossil fuels is having measurable impacts on the Earth's climate, with even more profound impacts likely in the future (Hansen et al. 2005a, Houghton et al. 2001, Karl and Trenberth 2003, Prentice et al. 2000). Moreover, fossil fuels are a non-renewable resource with uncertain future prices and availability owing to limited supplies and fragile international trade agreements. Thus, academic, industrial, and governmental researchers are increasingly exploring renewable sources of energy.

Potential sources of sustainable energy include solar, geothermal, hydroelectric, wind, and biomass. Although each of these options has positive and negative attributes, biomass energy holds immediate promise because it is broadly available, fairly well developed technologically, and in some cases can be linked to other benefit streams in addition to the production of energy. In the United States, interest in woody biomass as a fuel is increasing as both an alternative fuel and a means of reducing fire risk near forested communities (GAO 2005).

The two primary obstacles that currently limit the use of woody biomass in the United States are low cost-effectiveness and lack of reliable supply (GAO 2005). For example, the cost of producing electricity from woody biomass using current technologies in the United States is currently 7.5 cents per kWh, whereas the market price for this electricity is only 5.3 cents per kWh (GAO 2005).

These obstacles might be overcome if selected communities can institute pilot projects that demonstrate the efficacy of biomass energy, provide a testing ground for improvements, and at the same time enjoy immediate economic and social benefits locally. We propose that the ideal locations for such pilot projects might be in communities with the following attributes:

- Relatively small and self-contained with simple infrastructure
- High current cost of power and/or heat
- Proximity to sustainable supplies of woody biomass
- Lack of social opposition to use of biomass fuel
- Strong social impetus to mitigate global climate change
- Interest in obtaining marketable carbon credits
- Existence of other social and economic considerations that make biomass harvest and use a desirable option.

Many villages and towns in interior Alaska fit all of these criteria. Rural Alaskans are disproportionately exposed to the effects of climate change, which is most pronounced at high latitudes (ACIA 2005), and struggle with rising fuel costs in a mixed economy characterized by high transportation costs. In rural Alaskan communities, mainstream fossil fuel technologies are prohibitively expensive. Large quantities of alternative fuels in the form of woody biomass (chiefly black spruce, *Picea mariana* (Mill.) B.S.P.) are available in this region, and the technology to use these fuels is relatively simple. Moreover, positive economic externalities may be realized through forest thinning or clearing, given the risks of forest fires to life and property, the direct costs of fire suppression, and the negative impacts of fire suppression on long-term ecosystem services. The advent of carbon trading markets in both the public and private sectors provides a source of additional revenue for alternative energy projects that could potentially tip the balance toward renewable energy sources (Duval 2004), although because such markets are slow to develop, this analysis does not depend upon their existence. Biomass can be used for heating, for energy generation, or for combined heat and power. This paper's focus is at the village level rather than the household level at which many heating choices are made; thus we chose to explore the possibility of conversion of village diesel generation facilities to renewable energy sources as one way in which villages might partially mitigate climate change, earn tradable carbon credits, reduce fuel costs, reduce fire risk, and increase local autonomy, thereby reducing vulnerability to external social and economic change.

In many regions both in the United States and abroad, immediate transition to alternate fuels is limited for economic, technological, or sociopolitical reasons. However, in much of interior Alaska, economic drivers, governmental infrastructure, available natural resources, and social imperatives all point toward the viability of conversion to new energy sources. We suggest that fuel conversion programs could be implemented in such a manner as to have positive effects on these systems. We further suggest that interior Alaska has the opportunity to provide leadership in this arena.

Previous studies have examined the feasibility of using wood fuel for energy generation in particular communities, including Dot Lake (AEA 2000a) and McGrath (Crimp and Adamian 2001). However, these studies cannot easily be extrapolated to other communities, and do not examine such factors as fire risk reduction and job creation. In this paper we provide a more comprehensive assessment. We analyze the feasibility and sustainability of potential biomass energy programs in rural Alaska by creating a social, biological, and political model framework within which we evaluate not only a wider range of financial costs and benefits, but also the interactions of ecological feasibility, social acceptability, community interest, and leadership commitment.

BACKGROUND: SYSTEM COMPONENTS

Energy Systems in Rural Alaska

Approximately 200 villages in Alaska have no connection to the electrical grid that links Alaska's largest communities. Prior to the 1960s, electricity was not available to most rural Alaskans (AVEC 2005). Now, these villages are generally supplied with electricity by diesel generators ranging from about 15 to 3100 kW in energy output (AEA 2000b). In total, 382 971 145 kWh of power were produced through diesel generation in Alaska in 2004, and 28,476,898 gal (107 459 992 L) of diesel fuel were consumed (AEA 2004). Many rural communities are part of regional cooperatives, including the

Alaska Village Electrical Cooperative, Inc. (AVEC), which operates more than 150 diesel generators in 51 communities that run a cumulative 414,822 hours a year (AVEC 2005).

Because most rural Alaskan communities are not on the road system, fuel for these generators must be transported by barge or airplane. Thus, in most cases, fuel can only be transported during summer, and enough fuel to last a full year must be stored on site (Colt et al. 2003). Maintaining this large storage capacity for fuel has posed significant environmental problems and incurred hundreds of million dollars of expenses (Colt et al. 2003, Duval 2004).

Because of the high costs of fuel transport and storage in rural Alaska, energy prices are extremely high. Consumers pay an effective rate of up to 35 cents per kWh in some regions. Less than half the total cost of electricity in rural Alaska can be directly attributed to fuel costs (Colt et al. 2003). Storage alone adds an estimated $0.40/L, owing to capital expenses and spill response capability—which itself may add as much as $0.16/L (UAF 2005).

Even in urban areas, electricity is more expensive in Alaska than in other parts of the country. In Fairbanks, the largest community in the interior and Alaska's second-largest city, residential power costs over 11.6 cents per kWh, not counting additional charges (GVEA 2005), 35 percent more than the nationwide average cost of residential electricity (EIA 2005).

In rural areas, much higher costs occur despite substantial subsidies. Alaska's Power Cost Equalization Program (PCE) provides assistance based on an algorithm that discounts costs between 12.0 and 52.5 cents per kWh by 95 percent (AEA 2004). Average residential rates without the subsidy would be more than 60 cents per kWh in some communities. Even so, the combined costs borne by consumers and the PCE program still do not account for a large proportion of the real costs of the system, which are funded by government grants, mostly for infrastructure. For small independent villages that are not AVEC members, these grants cover more than half (55 percent) of the real costs; for AVEC members, they cover approximately 26 percent (Colt et al. 2003). As the umbrella group for all village energy programs, the Alaska Energy Authority (AEA) administers and/or funds rural power system upgrades, the PCE program, energy conservation and alternative energy development, circuit rider maintenance and emergency response, utility operator training, a bulk fuel revolving loan fund, a power project loan fund, and maintenance of AEA-owned facilities. Although AEA has its own capital fund, recent capital project funding for bulk fuel storage upgrades and rural power system upgrades has come primarily from the Denali Commission, a federal-state partnership established by Congress in 1998 to provide critical utilities, infrastructure, and economic support throughout Alaska. It has been supplemented by other federal grants from agencies such as the Environmental Protection Agency (EPA) and the Department of Housing and Urban Development (HUD), as well as by state appropriations for capital expenditures.

Rising fuel price is likely to be the single greatest driver for a change from diesel-only systems. Diesel power generation is expensive in both direct and hidden costs. Among these are air pollution; problems with effective storage, resulting in soil and groundwater contamination from spills; spills during transport or transfer, resulting in larger scale contamination and risks to humans and wildlife; risk of nondelivery of fuel under adverse conditions, resulting in loss of power; and dependency on the PCE program (Colt et al. 2003). A typical rural village has separate tank farms owned and operated by the city government, the tribal government, the village corporation, the local school, the electric utility, and other public or private entities. As of 1999, the EPA considered 97 percent of these tank farms to have serious deficiencies, including inadequate foundations, dikes, joints, and piping; improper siting near water sources; and rust and corrosion (EPA 1999, Poe 2002).

Biomass Investment and Technology

Developing village biomass projects is timely, given new interest and potential funding for wood energy in interior Alaska. The Alaska Wood Energy Development Task Group, a recently formed coalition of federal and state agencies and other not-for-profit organizations, is now actively coordinating the state's efforts to increase the use of biomass for energy in Alaska. Since 2004, the task group has been soliciting biomass energy project proposals from communities for funding with AEA-earmarked funding. As of 2007, AEA had budgeted $669,674 for wood energy activities (AEA 2005, AEA 2007).

Wood fuel has traditionally been converted into energy via open burning, fireplaces, and wood stoves. In traditional applications, the energy efficiency of biomass fuels for heating, cooking, and energy production is very low—in some cases as low as 10 percent (Kishore et al. 2004). However, biomass technology has improved over the past decade and has enjoyed success in other parts of the world, including Scandinavia and India. New biomass technologies allow for both more efficient energy conversion and—owing to a hotter and more complete burn—greatly reduced emissions of particulates and carbon monoxide. Biomass fuels can include whole trees, cut firewood, chunk-wood, compressed sawdust pellets or briquettes, or gasified wood. These fuels can be used for electricity generation, heating, or a combination of both. Modern methods that offer greater combustion efficiency and lower emissions of air pollutants include combustion in a modern boiler/steam turbine system, direct wood gasification, or pyrolysis (Bain et al. 1996). Although energy release is highly efficient in all of these systems, considerable energy is lost in converting that energy to electricity. Typically, the overall efficiency of a system that is only used to generate electricity is a mere 25 to 30 percent (Bain et al. 2003). However, much of the energy lost is converted to heat. If heat is also a desirable product, as is the case for most of the year in interior Alaska, the boiler system can be configured for the simultaneous production of heat and electricity. More than 50 rural Alaska communities—or approximately 27 percent—already have combined heat and power (CHP) systems (Crimp and Adamian 2001, MAFA 2004) and therefore have the infrastructure for heat and power distribution. Although system configurations range widely, a preliminary assessment of the market indicates that 70 percent of rural Alaska communities could make cost-effective use of combined heat and power systems (MAFA 2004).

Boiler systems are the simplest choice for biomass heat and power generation. In such a system, whole-tree wood chips or chunks are oxidized with excess air circulation, either in a stoker or a fluidized bed, and the hot flue gases released produce steam in the heat-exchange sections of a boiler. Some of this steam produces electricity via a turbine in a Rankine cycle, and the excess steam is used for heat (Bain et al. 2003).

Wood gasification and pyrolysis are potentially 30 to 40 percent more efficient than direct combustion, require less water, and result in cheaper costs per kWh, but generally involve more complex operation and maintenance requirements and newer and less proven technology.[1] Wood gasification is the process of heating wood in an oxygen-limited chamber to a temperature range of 200 to 280 °C until volatile gases including carbon monoxide, hydrogen, and oxygen are released from the wood and combusted (Bain et al. 2003). Several methods of gasification exist; however, updraft gasifiers are the

---

[1] Scahill, J. 2003. Biomass to energy: present commercial strategies and future options. Presentation. Denver, CO. Healthy Landscapes and Thriving Communities: Bioenergy and Wood Products Conference. U.S. Department of the Interior. Jan. 21.

simplest and most reliable (see footnote 1) and thus the only type considered in this analysis.

Carbon Markets

Although the United States is not a signatory to the Kyoto Protocol on Climate Change, and policy analysts predict that carbon dioxide (CO_2) reductions will not become mandatory in the United States in the near future (McNamara 2004), the ramifications of this international agreement, as well as the dialogue that led to its creation, have nonetheless altered the way in which U.S. carbon stocks and fluxes are likely to be managed in the future.

In signatory nations, long-term carbon sequestration has become a commodity that can be traded against carbon emissions based on a cap-and-trade system (McNamara 2004). Likewise, reduction of emissions from nonrenewable sources (generally fossil fuels) can be traded against increases in other sectors. In January 2005, the European Union—including all 25 of its member states—initiated the European Union Emissions Trading Scheme (ETS), a legally binding international trading market in greenhouse gas emissions. Russia, Canada, and Switzerland are working toward instituting parallel systems (Kirk 2004). The transferability of carbon credits has opened up international economic possibilities never before seen, although some parallels can be drawn to the successful mitigation of sulfur dioxide pollution in the United States through use of tradable pollution credits (CCX 2006).

Meanwhile, nongovernmental markets have already appeared, even in non-signatory nations. In the United States, the Chicago Climate Exchange (CCX) is currently the most viable carbon credit market (McNamara 2004). It is acting as a self-regulating voluntary market, administering the world's first multisector and multinational emission-trading platform. By participating in trading through CCX, corporations, municipalities, and other institutions have made legally binding commitments to reduce net emissions of greenhouse gases. Carbon emitters as well as credit holders are banking on future increases in the price of credits because of either international agreements or state and local laws. By entering the market early, buyers are showing good will and environmental responsibility, as well as setting up relationships that may prove lucrative in the future (McNamara 2004).

Alaska has yet to participate in nascent carbon markets, although the passage into law of a bill promoting carbon credit research (Berkowitz 2004) demonstrates the state's interest in both climate change and carbon-credit trading. Some states and geographic regions are already making local commitments to reduce greenhouse emissions. For example, in August 2001, the New England Governors and Eastern Canadian Premiers signed a regional climate change agreement aimed at reducing greenhouse gas emissions to 1990 levels by 2010, and reducing emissions to 10 percent below 1990 levels by 2020. To meet the requirements of this agreement, participatory states are creating local control mechanisms. In California, Governor Schwarzenegger signed Executive Order S-3-05 in June 2005, dictating that the state's greenhouse gas emissions would be reduced to 2000 levels by 2010, to 1990 levels by 2020, and to 80 percent of 1990 levels by 2050 (Schwarzenegger 2005).

Under the rules of the Kyoto Protocol—which are often used as guidelines, even in nonsignatory markets—tradable credits can be obtained in a number of ways, including afforestation, reforestation, and conversion from fossil fuel use to carbon-neutral fuels. For the purposes of carbon accounting, biomass can be considered carbon neutral: although carbon is emitted when biomass is burned, forest regrowth should, over time, take up an equal quantity of carbon. However, because the time scales of emissions and absorption differ, the sustainability of the forests from which biomass is harvested must be certified. All emission reductions and tradable carbon credits must be monitored, verified, and certified by a third party that provides both confirmation that the carbon exists and insurance that it will be sequestered for the duration of the commitment period. Marketable carbon offsets also require proof of additionality—an assurance that sequestration or emission reductions would not have occurred had the project not been implemented. Finally, projects must not lead to "leakage": emission increases in another sector that can be attributed to reductions in the credited sector (Innes and Peterson 2001, UN 1997).

In interior Alaska, fuel substitution may hold the greatest promise for attaining marketable carbon credits. Unlike credits based on afforestation, reforestation, or increased forest stocking, fuel offset credits are not one-time credits; as more fossil fuel use is offset over time, more credits can be earned. In addition, biomass energy generation can theoretically be developed on a wide range of scales. Finally, as described above, fuel offsets may be possible within a framework that generates other positive outcomes in addition to reduction of carbon emissions.

Forest Ecology and Ecosystem Services

The ecological sustainability of any proposed biomass fuels project will be pertinent not only from the point of view of achieving certifiable forestry practices in order to verify carbon sequestration credits, but also from the perspective of maintaining other ecosystem services. Historically, naturally occurring fires in interior Alaska have created a variegated landscape with multiple age classes of forest succession (Dyrness et al. 1986), each of which provides different resources (e.g., berries, moose browse, cover for furbearing mammals, and habitat for woodland caribou). However, fire suppression around inhabited areas tends to decrease average annual area burned (Dewilde and Chapin 2006), which over time will tend to increase average forest stand age and reduce this variability while also increasing the risk of future fires. Although harvest and fire do not result in identical post-disturbance trajectories (Rees and Juday 2002), harvest does offer a means of introducing age-class variability and reducing fire risk around communities.

GOALS AND OBJECTIVES

The purpose of this study is to assess the feasibility of switching from fossil fuels to wood energy in rural Alaska villages located in forested regions of interior Alaska (fig. 1) that are not supplied with electricity via the railbelt (the centralized power grid connecting Anchorage, Fairbanks, and other relatively large communities). More specifically, the study's objectives were to:

1) Create a quantitative ecological model of the footprint of potential biomass harvest for wood energy around interior Alaska villages.
2) Create a quantitative economic model of the short- and long-term costs and benefits of switching from diesel energy to wood energy in these remote communities.
3) Explore the effects of model input selection and model parameter uncertainty on model outputs.
4) Qualitatively assess the effects of social factors on the feasibility of fuel substitution.
5) Examine potential feedback between ecological, economic, and social factors, and assess ways in which they might in combination affect the feasibility of wood biomass fuel use in Alaska villages.

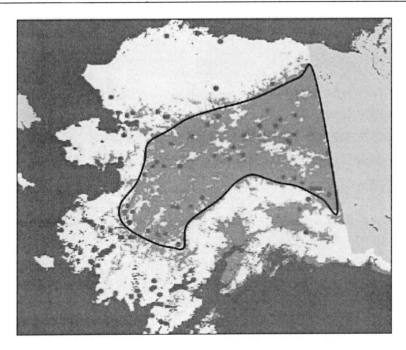

Figure 1. Remote Alaska communities. About 90 communities (represented by dots) lie in forested regions (green-shaded area). Approximately half of these are in the Interior region considered in this study (roughly demarcated by black line). Adapted from Crimp and Adamian 2000.

METHODS

Ecological Feasibility

For selected interior Alaskan villages, we created a simple model to estimate the area required to supply aboveground tree biomass over a rotation length that would mimic natural fire cycles while reducing fire risk in communities, optimizing aesthetic and subsistence values, and protecting ecosystem integrity. The biomass required was calculated from input variables and model parameters selected based on published data. Input variables included village size, village per capita energy needs, and optimal harvest rotation length. Parameters internal to the model included forest cover, forest volume, predicted biomass growth curves, and energy outputs by harvest volume.

Model output was expressed as maximum travel distance to obtain wood fuel—in other words, the distance between a village and the perimeter of the circle circumscribing the area of sustainable yield necessary to meet the needs described by the input variables. The radius (r) of a circle of area A is defined as

$$r = \sqrt{\frac{A}{\pi}}$$

The area (A) necessary for fuel collection around a village would be a function of the population and its energy needs, the percentage of those needs to be met by biomass, the percentage of land included as productive for black spruce, the energy available per acre of wood harvested, and the frequency with which any particular acre could be harvested. Thus, the general formula used was

$$D\,max = \sqrt{\frac{P \times Epc \times Eo \times R \times 0.01}{Bd \times Ad \times Ew \times Ee \times Fc \times \pi}}$$

Where:

$Dmax$ = maximum travel distance (km)

P = village population

Epc = per capita energy use (kWh/yr)

Eo = Energy offset (fraction of total energy use replaced with biomass energy)

R = Rotation length for forest harvest (years)

Bd = biomass density (t/ha) for black spruce at harvest age (green weight)

Ad = correction factor for converting green to air-dried wood (t air-dry/t green) Ew = energy available from air-dried wood (kW/t)

Ee = electrical efficiency (fraction of gross heating value converted to electrical energy)

Fc = Forest cover (black spruce forest as fraction of total land area)

0.01 = the correction factor to convert from hectares to square kilometers

We first obtained model results for villages within the study area by using mean, median, or generally accepted values as initial model parameters, hereafter referred to as "nominal" values. Nominal parameter values were selected conservatively, so as to overestimate rather than underestimate the footprint of harvest for biomass fuels around any particular village. Likewise, parameter ranges were selected to represent a relatively broad set of possible outcomes. Because all model inputs and parameters were part of a single first-order equation, and because all variables were multiplicative, the sensitivity of the model to variability in each parameter depended only on the magnitude of the range of possible values for that parameter. However, some of these ranges were quite large, resulting in a substantial cumulative effect of parameter uncertainty. We examined the sensitivity of the model to uncertainty in both model inputs and model parameters by performing 300 stochastic model runs—100 each for minimum, mean, and maximum community sizes—using parameter values randomly selected from within each parameter range.

Model inputs reflected known or predicted values for village sizes and energy usage based on Alaska census data and information published by the AEA (AEA 2000b, 2002, 2004; ADCED 2005) (table 1). Mean population for the communities we focused on was 106, with a range from 21 to 1,439. We considered energy use at current levels, based on kWh generated rather than kWh actually used in order to account for inevitable waste. The mean value was 3758 kWh per capita, close to the 4000 kWh estimated by Colt et al. (2003). Communities with the highest usage were similar to the U.S. average of 10,000 kWh per capita (Colt et al. 2003).

Rotation length was also treated as a model input, as it depends on community preference. We assumed that communities would seek to reduce wildfire risk as a byproduct of their harvest strategy and that they would therefore only harvest mature black spruce stands (the most fire-prone landscape type). An 80-year rotation would allow for harvest in early maturity, whereas a 200-year rotation would yield trees in late senescence; very few stands older than 200 years can be found for any species in interior Alaska (Yarie and Billings 2002). Thus, we bounded the range of inputs with these values. The nominal value was set at 110, just prior to apparent age- and/or fire-related decreases in stand frequency (Hollingsworth 2004, Yarie and Billings 2002).

Table 1. Energy use and costs in forested interior Alaska communities not on the railbelt electrical grid

Community	Population[a]	Per capita Installed rate residential Electrical use[b]	Fuel use[c]	Average price[b]	Installed generator capacity	Residential rate without PCE[b]	Actual residential rate w/PCE[b]
		kWh	*Gallons*	*Dollars per gallon*	*kW*	*Dollars per kWh*	*Dollars per kWh*
Alatna and Allakaket	122	5318	53,773	2.19	430	0.48	0.27
Aniak	532	4640	192,576	1.32	2865	0.49	0.32
Anvik	101	4644	38,474	1.32	337	0.46	0.28
Beaver	67	4379	31,436	1.92	137	0.42	0.26
Evansville and Bettles	51	13 800	58,368	1.41	650	0.41	0.20
Central	102	4921	50,104	1.22	640	0.51	0.28
Chuathbaluk	105	2036	20,200	1.70	n/a	0.56	0.32
Circle	99	3758	34,750	1.24	200	0.50	0.27
Crooked Creek	147	1731	25,258	1.69	n/a	0.56	0.32
Dot Lake	29		n/a	n/a	325	0.23	0.17
Eagle and Eagle Village	183	4270	58,474	1.20	477	0.41	0.26
Fort Yukon	594	4781	207,698	1.66	2400	0.34	0.23
Galena	717	13 203	724,076	1.46	6000	0.25	0.18
Grayling	182	3235	46,352	1.52	546	0.44	0.28
Healy Lake	34	4500	14,339	1.25	105	0.40	0.24

Community	Population[a]	Per capita Installed rate residential Electrical use[b]	Fuel use[c]	Average price[b]	Installed generator capacity	Residential rate without PCE[b]	Actual residential rate w/PCE[b]
Holy Cross	206	3437	54,340	1.51	585	0.42	0.27
Hughes	72		37,325	3.27	323	0.51	0.30
Huslia	269	3409	77,648	1.79	680	0.46	0.28
Kaltag	211	3143	57,498	1.58	573	0.46	0.28
Koyukuk	109	3241	20,830	1.89	244	0.45	0.36
Lime Village	34	2920	9,101	4.44	77	0.80	0.56
Manley Hot Springs	73	4029	26,772	1.14	480	0.60	0.36
McGrath	367	8074	221,650	1.40	2685	0.43	0.29
Minto	207	3491	56,366	1.13	558	0.40	0.26
Nikolai	121	3317	38,182	1.81	362	0.50	0.34
Northway and Northway Village	195	8123	121,569	1.29	1165	0.43	0.25
Nulato	320	3590	85,982	1.59	897	0.44	0.28
Red Devil	35	3612	14,490	1.83	173	0.56	0.32
Ruby	190		24,861	1.76	654	0.46	0.33
Shageluk	132	3073	31,506	1.69	370	0.46	0.28
Sleetmute	78	2939	25,314	1.69	208	0.56	0.32
Stony River	54	2156	13,994	1.69	139	0.56	0.32
Takotna	47	5292	28,219	1.72	297	0.48	0.32

Table 1. (Continued)

Community	Population[a]	Per capita Installed rate residential Electrical use[b]	Fuel use[c]	Average price[b]	Installed generator capacity	Residential rate without PCE[b]	Actual residential rate w/PCE[b]
Tanana	304	4533	104,270	1.34	1456	0.49	0.31
Tetlin	129	3669	40,782	1.46	280	0.47	0.27
Tok	1,439	8700	861,311	1.25	4960	0.23	0.17

Note: The penultimate column indicates what electrical rates would be in each community if Power Cost Equalization (PCE) subsidies were not provided by the state, and the final column shows the actual rates paid by householders.

n/a = not available.

[a] Data from ADCED 2005.

[b] Data from AEA 2004.

[c] Data from UAA 2003.

Across the interior, black spruce stands account for approximately 44 percent of the landscape (Sharratt 1997). This was used as a nominal value, although the actual mean is likely to be higher because of undercounting of early-succession stands that would be classified as black spruce in a later successional stage. Because villages in areas with less than 10 percent forest cover were not considered, 10 percent was set as the low value, and 75 percent was selected as an upper limit (Fitzsimmons 2003). Although forest cover approaches 100 percent in some regions of the interior, land around villages often contains considerable areas of rivers and other wetlands, so a conservative estimate was chosen.

Table 2. The heating value of wood

Moisture content	Gross heating value		
	Low	Medium	High
Percent	*kWh/t*		
0	5025	5490	5761
25	3769	4118	4321
30	3518	3843	4033
35	3266	3569	3745
40	3015	3294	3457
45	2764	3020	3169
50	2513	2745	2881
55	2261	2471	2593
60	2010	2196	2305

Note: Values for a wide selection of hardwoods, softwoods, and wood residues fall in a relatively narrow range, with black spruce near the high end. Gross heating value depends primarily on moisture content.

The energy value of dry spruce chips was bracketed within a relatively small range by different authors (Maker 2004, Somashekhar et al. 2000, Zerbin 1984), making our model relatively insensitive to changes in this parameter. Based on these estimates, we selected a nominal value of 8,500 btu/lb (5480 kWh/t), with low and high boundaries of 7,780 and 8,920 btu/lb (5018 and 5753 kWh/t). However, differences in moisture content substantially affect energy output, because in the case of high-moisture fuel, some of the energy released by combustion is used to evaporate water (table 2). Although many wood burner systems can be used with a wide range of fuel types and fuel moistures, air-dry black spruce was selected as the nominal fuel, owing to the general availability of the species and the relative technological ease of air-drying as compared to kiln-drying.

Green black spruce has a moisture content (MC) of approximately 60 percent (Yarie and Mead 1982), whereas air-dried wood has approximately 12 to 15 percent moisture (Prestemon 1998, Yarie and Mead 1982). Although this figure may in some cases be lower in Alaska's dry climate, we assumed an air-dried moisture content of 15 percent, and thus a typical weight loss of 28 percent during the drying process, and a final gross heating value (GHV) of 85 percent of the oven-dry value. Boundary values for these parameters were set at 0 percent weight loss and 40 percent GHV for green wood (table 2), and 31 percent weight loss and 90 percent GHV for wood at 10 percent moisture.

Table 3. Electrical and total efficiency of wood-fired systems

Type of process	Electrical efficiency	Combined heat and power efficiency	Source
Hot gasification/fuel cell	0.23	0.6	Osmosun et al. 2004
Downdraft gasification	0.40	0.9	Zerbin 1984
Gasification		0.7	Wu et al. 2003
Gasification	0.35		Willeboer 1998
Gasification/fuel cell	0.24	0.6	McIlveen-Wright et al. 2003
Combustion	0.25		USDA 2004
Biomass integrated gasification combined-	0.33		Haq 2002
Gasification	0.21		Somashekhar et al. 2000
Combustion	0.20	0.6	Bain et al. 2003
Mean	0.28	0.68	

Note: Most authors report greater efficiency from gasification systems than from direct combustion.

Average aboveground tree biomass (including the fresh weights of bole, branches, and foliage) for 80-, 110-, and 200-year-old black spruce stands in interior Alaska are approximately 25, 28, and 10 t/ha, respectively (Yarie and Billings 2002). It is likely that the low value for 200-year-old stands reflects the result of slow growth on shallow saturated soils; such stands would be less than optimal for biomass fuel management. We selected 28 t/ha as both the nominal and the maximum value, and 10 t/ha as the minimum value.

We assumed a nominal efficiency of 28 percent for electrical production, with a range of 20 to 40 percent, based on the estimates shown in Table 3. Overall efficiencies for combined heat and power systems are significantly higher. However, we chose to focus on the feasibility of wood-fired electrical generation and thus treated heat energy as a positive externality.

Economic Feasibility

Rural Alaskan villages have mixed economies that include significant market and nonmarket components, and the costs of current village energy programs are borne not only by community members but also by external entities. Thus, in order to analyze the economic sustainability of potential fuel offset programs, we considered not only the costs and benefits of construction, operation, maintenance, fuel, employment, and carbon sequestration credits for diesel versus biomass systems, but also circulation of cash income and noncash commodities within communities, and the effects of subsidies. We examined economic feasibility based on published estimates and projections for:

- Village energy consumption
- Fossil fuel cost
- Nonfuel expenses specific to diesel systems
- Existing subsidies for fossil fuels, infrastructure, and maintenance
- Installation and maintenance costs for biomass systems
- Labor and mechanical costs for wood procurement
- Existing village economies, cash flows, and employment
- Current and potential future prices for carbon credits

We created a quantitative model incorporating the above components to assess whether fuel conversion would be likely to have a positive economic outcome for each village, and over what period initial investments in biomass infrastructure might be recouped.

The model input was the biomass generation capacity installed. Parameters internal to the model included diesel prices, nonfuel expenses specific to diesel systems, nonfuel expenses common to both systems, installed diesel capacity, actual kWh of power generated, installation costs for biomass systems, and annual operation costs for biomass systems. For each of these parameters we either used published village-by-village values or determined nominal values based on mean, median, or generally accepted values from the literature. Nominal parameter values were selected conservatively, so as to overestimate rather than underestimate the costs of fuel conversion. Likewise, parameter ranges were selected to represent a relatively broad set of possible outcomes.

We examined the sensitivity of the model by randomly selecting parameter values for key variables (diesel price, biomass system installation costs, annual biomass operation and maintenance costs, and carbon credit prices) from the full range of uncertainty expressed in the literature. Using these random values, we analyzed the results of 10 stochastic model runs for each of the 31 villages for which adequate data were available.

The general formula used in the economic submodel was:

$$Y = \frac{CapitolCosts}{AnnualSavings} = \frac{CapitolCosts}{AnnualCostsOffset - AnnualBiomassCosts + AnnualCarbonCreditValue} = \frac{Ic \times El}{(Ao \times De \times Dp) + (NFo \times NFc) - (Bg \times Bc) + (De \times Ao \times Cc)}$$

Where:
Y = years to pay back investment
Ic = installed cost of a biomass power system, per kW generation capacity
El = electrical load (total biomass capacity installed, in kW)
Ao = actual offset, in kWh (based on relationship between installed biomass
capacity and mean electrical load)
De = diesel efficiency (gallons of diesel fuel per kWh generated)
Dp = diesel price ($/gallon for diesel fuel)
NFo = estimated nonfuel offset (fraction of nonfuel costs, e.g., fuel storage and spill prevention, offset by use of biomass)
NFc = total nonfuel costs (including diesel-specific costs and those common to biomass or diesel systems) (total $)
Bg = biomass energy generated (kWh/yr) Bc = Biomass energy costs ($/kWh)
Cc = carbon credits available owing to fuel offset ($/gallon fuel)

Total capacity installed in each village, total annual energy use in each village, and much of the data on nonfuel costs and existing costs and funding sources for power systems was available through state Department of Community and Economic Development budget requests (Poe 2001, 2002) budget reports (Alaska 2001, 2002), the University of Alaska Anchorage Institute of Social and Economic Research (UAA 2003), and the AEA (AEA 2000b, 2002, 2004, 2005). For the most part, these parameters were incorporated in the model as given. However, the proportion of nonfuel costs incurred prior to or during generation (e.g., the costs of fuel storage and boiler operation and maintenance) were not always separated from those incurred after generation (e.g., the costs of distribution and customer service). This breakdown had to be estimated based on partial data. Average fuel prices were based on 2004 figures, despite the steep rise in prices over the following years. However, we assessed the sensitivity of this parameter within the range of -50 to +150 percent to account for this volatility.

As a nominal model input, we assigned biomass capacity installed in each village a value equal to the mean electrical load for that community. Under this assumption, existing diesel systems would be at least partially retained and maintained to meet peak loads, while allowing biomass systems to run at full capacity for much of the time. In the communities we assessed, mean load was only 8 to 29 percent of installed capacity (appendix), demonstrating overcapitalization that would probably not be necessary to replicate with biomass systems. Load profiles are not available for most rural Alaskan communities. However, available information from six villages of varying sizes shows combined daily and seasonal variation yielding peak loads that are approximately twice mean loads and threefold the minimum loads (Devine et al. 2005). Installation of biomass generation capacity greater than minimum loads would result in some unused capacity; at a capacity equal to mean loads unused capacity would be about 30 percent, and at a capacity equal to twice mean loads it would be approximately 60 percent (figure 2).

Diesel fuel costs would be directly offset according to the number of kilowatthours actually generated by the biomass system. Nonfuel expenses would be offset by the percentage of these costs associated only with diesel systems and by the total capacity replaced. Nonfuel generation expenses for diesel systems are steep because they include construction and maintenance of fuel tanks as well as spill response capabilities, although not all of these costs are currently internalized (Colt et al. 2003). We estimated that continuous operation of biomass systems at mean load levels would offset 60 percent of the village's diesel fuel use, but reduce nonfuel expenses associated with existing systems by only 25 percent. To assess the sensitivity of the model to our assumptions, we compared the results with a model run in which biomass generation capacity replaced only 50 percent of mean loads, replacing 40 percent of diesel fuel use and 10 percent of nonfuel expenses (Devine et al. 2005).

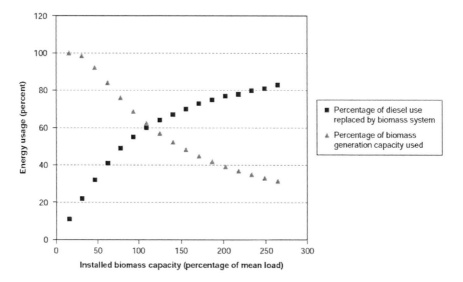

Figure 2. Biomass generation capacity and diesel fuel savings. Owing to daily and seasonal variability in energy demands, total system capacity is designed to greatly exceed average loads (adapted from data on substitution of diesel systems with wind power, Devine et al. 2005).

We compiled estimates of capital costs for purchase and installation of biomass systems from a range of available sources (table 4). To present conservative approximations in estimating feasibility of fuel conversion, and to allow for the potentially higher costs of installation and operation in remote Alaskan sites, we used the mean of the authors' high-end estimates, $1,849/kW, as the nominal value in our model. For the purposes of sensitivity analysis, we considered the range of values between the minimum published value ($980/kW) and 125 percent of the maximum published value ($2,500 x 1.25) = $3,125.

Generation costs (including fuel, operation, and maintenance costs) for wood-powered systems are difficult to accurately estimate, as they depend on location, wages, ease of fuel procurement, mechanization of harvest, and ease of maintenance. A national estimate of 7.5 cents/kWh (GAO 2005) seems far too optimistic for our purposes; in rural Alaska, travel costs and lack of local technical expertise would be expected to drive up the costs of system maintenance. However, this is already the case for diesel systems. Small-scale relatively nonmechanized methods for gathering and chipping wood might increase labor costs per ton of fuel, but the ready availability of both wood fuel and labor might partially balance these effects. We estimated generation costs based on actual costs of clearing and thinning projects in rural communities (table 5) (Hanson 2005, Lee 2005, USDI BLM 2005). In all cases, local crews were used, and the work was extremely labor-intensive and low-tech. Although these projects did not entail using the harvested wood for electrical generation, they did include manual disposal through piling and burning or chipping, as well as overhead and equipment costs. Translating these costs into equivalent energy costs resulted in a mean or nominal value of $0.16/kWh (rounded up to $0.17/kWh). To provide a more conservative estimate of feasibility in our sensitivity analysis and avoid reliance on a potentially anomalous value, we raised the lower end of this range to four times the costs recorded for Stevens Village (to $0.12/kWh), and rounded the upper limit to $0.28. Projected total generation costs (including fuel, operating, and maintenance) are similar to estimates of between $0.06 and $0.20/kWh (mean = $0.16/kWh) noted by various sources for small-scale or rural biomass projects (Bain et al. 2003, ENR 2001, Haq 2002, USDA FS 2004; see footnote 1) (table 4). This is substantially less than the real cost of diesel power in most villages, although it does not include the cost of distribution.

We gathered information on village-by-village fuel use, energy use, fuel costs, and subsidies primarily from annual statistical reports on the PCE Program (AEA 2000b, 2002, 2004) and Alaska Electric Power Statistics for 1960–2001 prepared by the Institute of Social and Economic Research at the University of Alaska Anchorage for the AEA, the Regulatory Commission of Alaska, and the Denali Commission (UAA 2003). Some of these data have already been shown in table 1; the full data set appears in the appendix.

We estimated model parameters for the value of carbon sequestration credits by gathering data on existing markets in the United States and Europe and calculating the tons of carbon offset for each 1,000 gal (3,774 L) of diesel replaced by biomass fuel. The estimated value of these credits covers a wide range, owing to market fluctuations and future uncertainty. Prior to the Kyoto Protocol taking effect in signatory nations, the trading price of carbon was typically slightly over $1 per metric ton. In 2005, prices fluctuated around the $2 mark, and we used a value of $1.90 in our analysis, despite the fact that more recent values have spiked as high as about $4.

Table 4. Capital costs and annual operation and maintenance (O&M) costs for biomass systems as compared to diesel generators

Estimated installed cost System			Estimated O&M costs					
type	Low	High	Low	High	Plant size	Plant type	Location	Source
Dollars per kW			*Dollars per kWh*		*kW*			
Biomass systems:	1,536	1,536	0.17	0.17	100,000	BIGGC[a]	U.S.	Haq 2002
	914	914	n/a	n/a	35,000	BIGGC	Brazil	Waldheim and Carpentieri 2001
	2,000	2,000	0.12	0.12	Up to 15	BIGGC	U.S.	ENR 2001
	1,230	1,488	n/a	n/a	5,000–10,000	FBC[b]	U.S., Finland	Bain et al 1996
	1,400	2,000	0.09	0.14	25,000–150,000	BIGGC	U.S.	Bain et al. 2003
	1,275	2,000	0.17	0.17	25,000–150,000	FBC	U.S.	Bain et al. 2003
	2,000	2,000	0.06	0.12	2,000–25,000	Unspecified	U.S.	USDA 2004
	980	2,500	0.15	0.20	1,000–110,000	GS[c] or FBC	U.S.	Scahil[d]

Estimated installed cost System			Estimated O&M costs					
type	Low	High	Low	High	Plant size	Plant type	Location	Source
Dollars per kW			*Dollars per kWh*		*kW*			
	900	2,200	0.15	0.20	15–650	BIGGC	U.S.	Scahill[d]
Mean	1,359	1,849	0.13	0.16				
Diesel generators (rural interior)	800	1,500	0.14	1.04	>100kW		U.S.	EIC 2002, AEA 2004

n/a = not available.

[a] BIGGC = Biomass integrated gasification combined cycle. Wood chips or chunks are heated in an oxygen-limited chamber to a temperature range of 200 to 280 °C until volatile gases including carbon monoxide, hydrogen, and oxygen are released and combusted.

[b] FBC = Fluidized bed combustion. Wood chips or chunks are directly combusted with excess air flow that circulates through the fuel bed.

[c] GS = Grate stoker. Wood chips or chunks are combusted in a simple stoker.

[d] Scahill, J. 2003. Biomass to energy: present commercial strategies and future options. Presentation. Denver, CO. Healthy Landscapes and Thriving Communities: Bioenergy and Wood Products Conference. U.S. Department of the Interior. Jan. 21.

Table 5. Costs per acre for forest clearing projects in rural Alaska villages

Fuel treatment project site	Type of treatment	Overhead and equipment cost per acre	Wages per acre	Total cost per acre	Cost per metric ton[a]	Operating cost per kWh[b]
		Dollars				
Healy Lake[c]	Fire break	640	2,560	3,200	282	0.22
Tanacross[c]	Parklike clearing to spacing of ~12 ft	800	3,200	4,000	353	0.27
Delta Junction[d]	Fire break	n/a	n/a	1,100	97	0.07
Stevens Village[c]	Light thinning of spruce understory	100	400	500	44	0.03
Fairbanks[e]	Fire break	n/a	n/a	2,700	238	0.18
Mean		513	2,053	2,300	203	0.16

Note: Costs vary depending on how labor-intensive the work is and how the project is managed. n/a = Not available.
[a] Assuming 28t/ha, 405ha/acre.
[b] Assuming 5,480 × 0.85 = 4658 kWh/t (green weight)
[c] Data from Hanson 2005.
[d] Data from USDI BLM 2005.
[e] Data from Lee 2005, hand-felling method only.

Table 6. Estimated annual quantity and value of potential carbon offset credits obtainable via fuel substitution in rural Alaska

	Diesel fuel		Carbon weight[c]	CO_2 emissions[d]	Value of carbon credits	
	Volume[z]	Weight[l]			CCXe	ECXI
	Liters	Kilograms	Kilograms	Tonnes	Dollars	Dollars
All PCE communities	107 796 786	84 081 493	72 049 266	263 700	501,031	6,328,807
Forested PCE communities in interior Alaska	13 329 974	10 397 380	8 909 494	32 609	61,957	782,610
Per 1,000 gallons of diesel	3785	2952	2530	9	18	222

CCX = Chicago Carbon Exchange, ECX = European Carbon Exchange, PCE = Power Cost Equalization Program.

[a] AEA 2004.

[b] Diesel fuel weighs approximately 0.78 kg/L.

[c] Diesel fuel is a mixture of hydrocarbons with an average weight ratio of 12 parts carbon to 2 parts hydrogen, with small amounts of other elements such as sulfur.

[d] When combusted, each carbon atom combines with two oxygen atoms at weight ratio of $C/CO2 = 3/11$.

[e] 2006 vintage, $1.90/t, September 2005 (CCX 2006).

[f] 20 €/t = $24/t August 2005 (McCrone 2005).

Although the international agreement had no direct effect on U.S. markets, it appears to have had an indirect effect (McNamara 2004). However, the prices of these voluntary credits remain far below the prices for verified emissions reductions in signatory nations. On the European Carbon Exchange (ECX), the European trading market, prices rose from approximately €8 ($9) at the beginning of 2005 to almost €30 ($38) in July 2005, and in August 2005 settled back down to about €20 ($24) (McCrone 2005).

Carbon credits represent a benefit stream from outside the village economy, with a value additive to all other benefit streams. We analyzed the potential value of the credits that could be obtained on a village-by-village basis, based on the number of tons of diesel offset, as determined by village energy use and biomass capacity installed (model input) (table 6). Although derived via different algorithms, our results, which estimate a total of 32,609 t of CO_2 emissions from diesel power generation in rural interior forested communities, are congruent with those obtained by Duval (2004), who estimated a total of 274 263 metric tons of CO_2 emissions for all PCE communities, with 52 047 of these tons from "forested Alaska." Our somewhat lower figures for forested interior Alaska reflect the fact that some rural forested communities are in the southeastern or south-central parts of the state, which are not considered in our analysis.

Social Feasibility

Analysis of social feasibility was primarily qualitative rather than quantitative, and included assessment of:

- Existing social infrastructure related to village electrical utility management and funding, fire prevention, and biomass harvest
- Threshold requirements (make-or-break factors needed within a particular community or at a broader scale, e.g., a minimum level of local technological expertise)
- Existing institutional barriers to change
- Potential positive social feedback (e.g., autonomy, employment)
- Potential negative social feedback (e.g., reactions to system quirks or failures)
- Lessons learned from existing biomass projects in rural Alaska

Although funding for village power systems is provided to a large degree by state and federal subsidies via AEA programs, ownership and operating responsibility for many of these projects is placed entirely with local grantees (Poe 2002). Thus, we assumed that most ultimate decisionmaking would take place at the village level, although financing, training, infrastructure, and technological expertise might all come from farther afield.

In addition, we drew information from past and ongoing projects with goals and objectives similar to those proposed in this study. These include wood fuel projects such as the existing boiler at Dot Lake and the proposed biomass system in McGrath (Adamian et al. 1998, AEA 2000a, Crimp 2005, Crimp and Adamian 2001); other alternative fuel projects such as wind-diesel hybrid systems (AEA 2005, Devine et al. 2005, MAFA 2004) and fire prevention efforts that include forest clearing (Hanson 2005, Putnam 2005).

Several fuel treatment projects aimed at reducing the risk of catastrophic wildfire have already taken place in village settings, under a combination of local leadership and assistance from entities such as the Alaska Department of Natural Resources (DNR) Division of Forestry and Tanana Chiefs Conference (TCC). The immediate costs of these projects were noted in table 5. However, to further ascertain the impacts of these efforts at the village level, we spoke with Doug Hanson of DNR (2005) and Will Putnam of TCC (2005). In particular, we questioned the importance of local hire; the role of key leaders, elders, or crew bosses; and the relationship between fire crews, harvest crews, and local opinions regarding fire protection.

Although for the purposes of the economic submodel we calculated costs and benefits irrespective of the impacts on different funding sources and beneficiaries, analysis of benefit streams was necessary for a more indepth understanding of the social submodels. Thus, we qualitatively assessed the current discrepancy between the real cost of power and the cost borne by consumers, the potential impacts of shifting funding and changing subsidies, and the potential economic value of local jobs generated by the harvest of biomass fuels. Our analysis was based on data on existing sources of funding for Alaska rural energy projects (table 7), data from the PCE Program (appendix) (AEA 2004); and financial information from past forest clearing projects (table 5)

Table 7. Annual funds for rural Alaska energy projects, including loans and grants

Funded item/activity	Federal funds (EPA, HUD, CDBG, DOE)	State appropriations	State revolving loan[a]	Alaska Energy Authority capital funds[b]	Denali Commission	Local funds	Unspecified	Total funding	Reference year
				Dollars					
Circuit rider maintenance and emergency response	100,000	200,000						300,000	2001
Utility operator training								n/a	
Rural power system upgrades							2,300,000	2,300,000	2000
Rural power operations	68,300	269,600					2,400,200	2,738,100	
Tank farm upgrades	4,900,000	2,450,000			15,350,000	550,000		23,250,000	2002
Bulk fuel revolving loan fund			51,000					51,000	2003
AEA power project loan fund			835,000					835,000	2003
Power cost equalization		15,617,225						15,617,225	2004
Energy cost reduction program[c]					2,500,000			2,500,000	2006
Village end use efficiency program[c]					722,000			722,000	2005

Funded item/activity	Federal funds (EPA, HUD, CDBG, DOE)	State appropriations	State revolving loan[a]	Alaska Energy Authority capital funds[b]	Denali Commission	Local funds	Unspecified	Total funding	Reference year
Wind energy assessment[c]	70,000			37,000	390,000			497,000	2005
Wood energy development program[c]	84,000			16,000				100,000	2005
Energy efficiency technical assistance[c]	137,500			62,500				200,000	2005
AEA operation and maintenance							1,067,100	1,067,100	2005
Total	5,359,800	18,536,825	886,000	115,500	18,962,000	550,000	5,767,300	50,177,425	

Note: All of these funds are managed by the Alaska Energy Authority (AEA).

EPA = Environmental Protection Agency, HUD = Housing and Urban Development, CDBG = Community Development Block Grant, DOE = Department of Energy.

[a] These funds are expressed as annual outlays. They are generally expected to be recouped and recirculated, but at zero or reduced interest rates.

[b] As of 2002, assets in the AEA fund were worth $800 million.

[c] Part of the energy conservation and alternative energy development program.

Data adapted from AEA 2005; 2002; 2004, Alaska Department of Community and Economic Development 2001, 2002.

RESULTS

Ecological Feasibility

Using nominal parameter values and a forest rotation length of 110 years, the maximum travel distance required to collect enough mature black spruce to meet average electrical loads (thus supplying approximately 60 percent of total village power) ranged from 1.1 to 12.8 km. (table 8).

Table 8. Estimated land area and maximum travel distance for sustainable harvest of black spruce for energy generation

Community	Population	Annual energy use	Load offset (biomass generation capacity = mean load)	Harvest area around village	Maximum travel distance
		kWh	*kWh*	*Hectares*	*Kilometers*
Alatna and Allakaket	122	648 861	389 317	2665	2.9
Aniak	532	2 468 700	1 481 220	10 140	5.7
Anvik	101	469 023	281 414	1927	2.5
Beaver	67	293 400	176 040	1205	2.0
Evansville and	51	703 820	422 292	2891	3.0
Central	102	501 896	301 138	2062	2.6
Chuathbaluk	105	213 737	128 242	878	1.7
Circle	99	372 000	223 200	1528	2.2
Crooked Creek	147	254 434	152 660	1045	1.8
Eagle and Eagle	183	781 344	468 806	3209	3.2
Fort Yukon	594	2 840 000	1 704 000	11 665	6.1
Galena	717	9 466 799	5 680 079	38 885	11.1
Grayling	182	588 761	353 257	2418	2.8
Healy Lake	34	152 986	91 792	628	1.4
Holy Cross	206	708 012	424 807	2908	3.0
Huslia	269	916 941	550 165	3766	3.5
Kaltag	211	663 172	397 903	2724	2.9
Koyukuk	109	353 250	211 950	1451	2.1
Lime Village	34	99 263	59 558	408	1.1
Manley Hot	73	294 120	176 472	1208	2.0
McGrath	367	2 963 200	1 777 920	12 171	6.2
Minto	207	722 562	433 537	2968	3.1
Nikolai	121	401 400	240 840	1649	2.3
Northway and	195	1 583 944	950 366	6506	4.6

Community	Population	Annual energy use	Load offset (biomass generation capacity = mean load)	Harvest area around village	Maximum travel distance
		kWh	kWh	Hectares	Kilometers
Nulato	320	1 148 831	689 299	4719	3.9
Red Devil	35	126 434	75 860	519	1.3
Shageluk	132	405 639	243 383	1666	2.3
Sleetmute	78	229 258	137 555	942	1.7
Stony River	54	116 418	69 851	478	1.2
Takotna	47	248 705	149 223	1022	1.8
Tanana	304	1 378 060	826 836	5660	4.2
Tetlin	129	473 310	283 986	1944	2.5
Tok	1439	12 518 973	7 511 384	51 421	12.8

With the exception of the two largest communities, Tok and Galena, which have regional and local road systems, respectively, the maximum travel distance was calculated to be 6.2 km or less, a distance easily reachable by snowmachine or four-wheeler, allowing for relatively low-tech harvest using chainsaws and a portable chipper. Larger communities might still find biomass fuel conversion an attractive option if they are located in regions with sufficient forest cover or road access, and if per capita electrical use remains modest. Even if 100 percent of village energy needs were supplied by biomass, the maximum travel distance for communities of up to 600 inhabitants would be no more than 8 km (fig. 3). Selecting a rotation length of 80 rather than 110 years only modestly reduces the maximum travel distance (fig. 4), because shorter rotations are correlated with lower biomass densities. However, increasing the rotation length to 200 years greatly increases the harvest area and travel distance, owing to both the longer return interval before stands can be harvested again, and reduced spruce biomass per hectare in older stands.

Model sensitivity analysis using randomly selected parameter values from within each parameter range yielded a distribution of results for each of three village sizes (fig. 5). For a village of 21 residents, no model runs yielded a maximum travel distance of over 3.8 km; the mean was 1.7 km. For a village of 106 residents, the range was 1.5 to 10.7, with a mean of 3.9 km. The distribution of results was broadest for the largest communities with a single outlier at 39.3 km. The remainder of the range fell between 5.5 and 27.5 km, with a mean of 14.2 km.

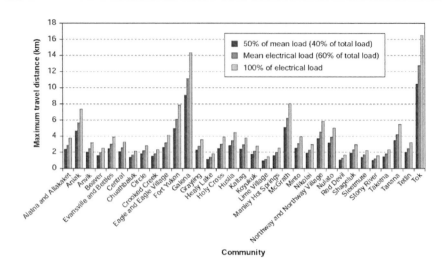

Figure 3. Maximum travel distance for meeting a given percentage of village energy needs by biomass fuels. Model outputs estimate sustainable harvest of black spruce for energy generation. If installed biomass generation capacity is equal to 50 percent of mean loads, approximately 40 percent of the community's electrical demand will be offset. At a capacity equal to mean loads, this rises to 60 percent of demand. All data assume 110-year forest rotations.

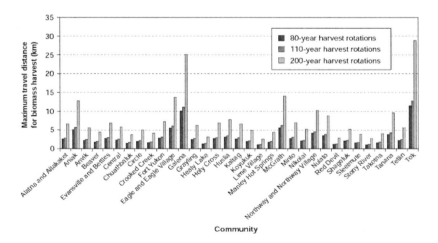

Figure 4. Maximum travel distance for sustainable harvest of black spruce for energy generation according to selected harvest rotations. Rotation lengths of 80, 110, or 200 years are shown. Black spruce biomass density per hectare increases between 80 and 110 years, and decreases between 110 and 200 years (Yarie and Billings 2002), resulting in a steep increase in travel distance with long rotations.

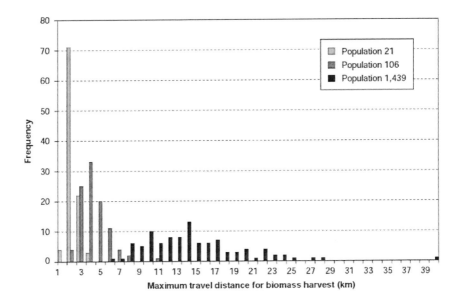

Figure 5. Distribution of results for 100 stochastic model runs for each of three village population sizes. In each model run, values for rotation length, biomass density, energy by moisture content, energy per ton, forest cover, and electrical efficiency were randomly selected from within broad accepted ranges.

Economic Feasibility

Because of missing data, not all economic calculations could be performed for all selected communities. For some villages, data were missing for fuel costs, nonfuel expenses, or energy generated (appendix), making it impossible to include these communities in model results. Thus, our results reflect a subset of forested off-grid villages in the interior. However, in addition to obtaining village-specific results, we were able to explore general relationships between village size, village accessibility, and economic feasibility.

For many of the communities in this analysis, total annual operating costs for electrical generation would be lower if part of the village's diesel power were converted to a biomass-fueled system (table 9). Tetlin, Tok, Northway, Koyukuk, Evansville and Bettles, and Eagle show consistently negative results; however, since Tok and Northway are both accessible via the Alaska Highway, one of the state's major thoroughfares, they may be considered anomalous as compared to more remote villages accessible only by minor roads or by rivers (major or minor) (appendix). Minto, Fort Yukon, and Tanana show benefits from conversion to wood fuel under some conditions but not under others, depending on the scale of the biomass generation capacity installed.

Table 9. Annual savings in generation costs and total capital investment associated with two levels of fuel system replacement

	Annual savings		Capital investment	
	Biomass capacity = 1/2 mean load	Biomass capacity = mean load	Capacity to meet ½ mean load	Capacity to meet mean load
			Dollars	
Alatna and Allakaket	11,320	25,317	68,479	136,957
Aniak	7,342	84,547	260,538	521,076
Anvik	146	11,945	49,499	98,998
Beaver	n/a	n/a	30,964	61,929
Evansville and Bettles	-7,444	-3,669	74,279	148,557
Central	5,176	22,618	52,968	105,937
Chuathbaluk	6,150	16,173	22,557	45,114
Circle	601	9,562	39,260	78,519
Crooked Creek	6,615	16,765	26,852	53,704
Dot Lake	n/a	n/a	n/a	n/a
Eagle and Eagle Village	-12,195	-5,423	82,460	164,921
Fort Yukon	-18,945	7,847	299,724	599,447
Galena	n/a	n/a	999,093	1,998,186
Grayling	2,865	19,017	62,136	124,272
Healy Lake	1,120	6,035	16,146	32,291
Holy Cross	2,377	21,266	74,721	149,442
Hughes	n/a	n/a	n/a	n/a
Huslia	16,168	47,175	96,771	193,542
Kaltag	7,822	28,313	69,989	139,978
Koyukuk	-6,399	-7,724	37,281	74,562
Lime Village	15,665	29,749	10,476	20,952
Manley Hot Springs	2,590	14,268	31,040	62,081
McGrath	-21,238	24,279	312,726	625,452
Minto	-5,593	9,675	76,257	152,513
Nikolai	4,549	11,024	42,362	84,725
Northway and Northway	-36,149	-45,395	167,164	334,328
Nulato	5,285	36,648	121,244	242,487
Red Devil	8,855	20,129	13,343	26,687
Ruby	n/a	n/a	n/a	n/a
Shageluk	3,856	15,924	42,810	85,619

	Annual savings		Capital investment	
	Biomass capacity = 1/2 mean load	Biomass capacity = mean load	Capacity to meet ½ mean load	Capacity to meet mean load
			Dollars	
Sleetmute	8,465	19,640	24,195	48,390
Stony River	8,450	19,582	12,286	24,573
Takotna	5,892	12,228	26,247	52,495
Tanana	-5,207	24,803	145,436	290,871
Tetlin	-4,680	-3,332	49,951	99,903
Tok	-353,480	-463,066	1,321,209	2,642,418
n/a = not available.				

When the added benefit stream of potential carbon sequestration credits is added to the potential annual savings gained by biomass fuel conversion, wood-fired electrical generation becomes more favorable. Even without taking carbon credits into account, 23 communities show a payback period of less than 25 years for the initial capital investment of installing a biomass electrical generation system adequate to meet mean electrical loads (fig. 6). A 25-year payback corresponds roughly to a real discount rate of 2.4 percent if the benefit stream continues for another 15 years. The projected time before a net positive economic balance is reached without carbon credits ranges from a mere 0.7 years for Lime Village and 1.3 years each for Stony River and Red Devil, to 11.7 years for Tanana and 15.8 years for Minto. If communities were able to sell carbon offset credits at 2005 U.S. prices, the payback periods for Tanana and Minto would drop to 11.3 and 14.9 years, respectively. At European carbon prices, these figures would dip to 7.8 and 9.2 years. Villages for which it would take longer than 25 years to recoup the investment and communities for which the benefit stream is negative are not shown in this figure. However, both McGrath and Fort Yukon, two of the larger communities analyzed, show a payback period of less than 25 years when carbon credits are taken into consideration, but not when carbon credits are not included.

It should be noted that, although villages for which data are absent have been necessarily omitted from this analysis, these communities should not be assumed to have a poor cost-benefit balance from potential biomass projects. In general, communities not accessible via a major road showed positive results based on biomass generation at mean load levels (fig. 7). This relationship was particularly robust for communities with fewer than 100 residents.

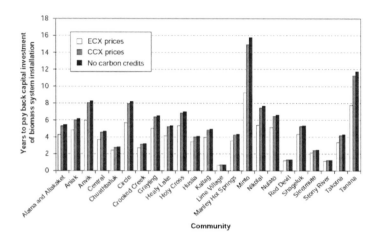

Figure 6. Years necessary to recoup an investment in wood-powered electrical generation capacity equal to mean electrical loads. For each selected village, three carbon-credit trading scenarios are shown: one in which no carbon credits are sold, one in which available fuel-offset credits are traded at Chicago Climate Exchange (CCX) prices, and one in which credits are traded at European Climate Exchange (ECX) prices.

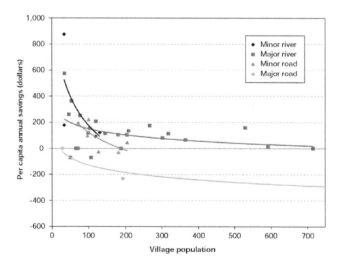

Figure 7. Per capita savings by village size and accessibility for biomass generation at mean loads. Logarithmic regression curves are fitted to four categories of accessibility. Only those villages that can be reached on major roads show consistently negative results for replacement of fossil fuel with biomass fuels at mean load capacity. For all villages, smaller population size is correlated with greater per capita benefits from fuel conversion.

For many communities, our model placed biomass conversion close to the economic break-even point when nominal parameters were used. Stochastic model runs using randomly selected parameter values from within broad possible ranges yielded mixed economic outcomes for almost all the villages analyzed (fig. 8). Only 10 villages—Aniak, Central, Chuathbaluk, Crooked Creek, Lime Village, Manley Hot Springs, Red Devil, Sleetmute, Stony River, and Takotna—showed net annual savings on generation costs for all 10 model runs. However, only two communities—Fort Yukon and Tok—yielded unfavorable results in 50 percent or more of model runs for replacement of mean load capacity. As the largest community with the greatest power usage, Tok also yielded the broadest range of potential annual costs or savings.

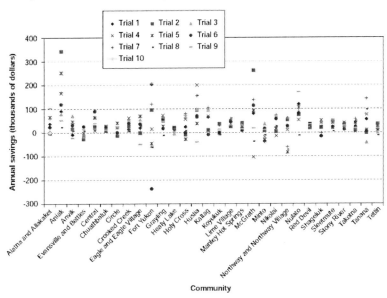

Figure 8. Sensitivity analysis for replacement of diesel systems with biomass electrical generation sufficient to meet peak loads. Data points show the results of 10 model runs using parameters randomly selected from within broad possible ranges. Tok has been excluded for reasons of scale; results for Tok ranged from -$1.9 million to +$1.5 million.

The ranges used in this analysis included installed costs between $980 and $3,125 per kW, annual operation and maintenance costs for biomass systems between $0.12 and $0.28 per kWh, carbon credits between $0 and $222 per 1,000 gal (3774 L) of fuel offset; and fuel prices between 50 and 250 percent of 2004 prices. It should be noted, however, that 2006 fuel prices were close to

200 percent of 2004 prices in many areas (DeMarban 2006b). If 2006 prices were used as a baseline, model runs would become consistently favorable in almost all communities.

When capital cost for biomass system installation was considered as a random stochastic variable and results were calculated for expected project payback time, results showed a similar pattern (fig. 9). Seven of the ten villages for which all model runs yielded annual savings showed a payback time of less than 25 years for all model runs. Only Tok showed a consistently poor ability to recoup the investment costs associated with biomass conversion, although other communities, including Evansville and Bettles, Fort Yukon, Koyukuk, Minto, and Northway yielded mixed results.

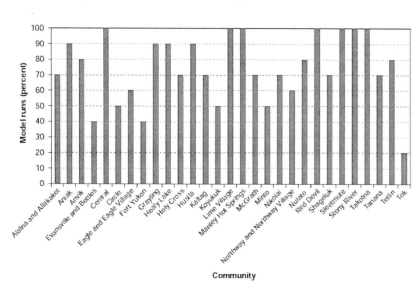

Figure 9. Sensitivity analysis of time necessary to recoup capital investment with biomass capacity of mean load. For each village, the graph illustrates the percentage of stochastic model runs for which randomly selected parameter values yielded a payback time of less than 25 years.

Social Feasibility

Our qualitative analysis of the potential social role of biomass fuel conversion in rural interior Alaska yielded a conceptual map of where wood fuel might fit into village economies (fig. 10). Harvest of biomass fuels would provide local jobs, which in turn would bolster the local cash economy by recirculating money within each village. In contrast, payments for fossil fuels represent a monetary flow out of communities. Currently, economic multipliers in village economies are small. Income from carbon credits would create a cash flow into the community from an outside source—something that is often in short supply in rural Alaska. Fire is linked to many aspects of community wealth, in both monetary and subsistence categories. Thus, natural forest succession, protection of life and property, local wages, and subsistence foods are all linked through the presence—or absence— of fire on the landscape.

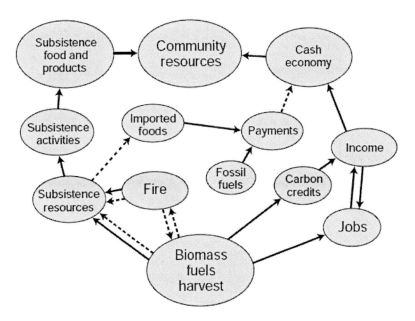

Figure 10. A conceptual model of economic feedback interactions. Village market and nonmarket economies are potentially linked to biomass fuels programs. Solid arrows indicate positive effects and dashed arrows indicate negative effects. Note that fire can have both positive and negative impacts on subsistence resources, depending on time scale.

Analysis of the impacts of subsidies and grants on village energy choices revealed a substantial gap between the real costs of electrical power and the prices being charged to consumers (fig. 11). Moreover, the real costs of village power make up a substantial proportion of village income, ranging from 7.1 to 70.0 percent (fig. 12). Because we have included the electricity used in shared facilities such as washeterias, schools, and offices in our totals, our figures are much higher than those for household use only (Colt et al. 2003). In reality, however, the discrepancy between realized costs and real costs may be even larger, owing to hidden (off-book) costs covered by transfer payments other than those made via the PCE Program. These include government-funded construction and upgrades, many of which were listed in table 7. Such off-book costs account for roughly 25 percent of the real cost of power (Colt et al. 2003), but are not accounted for in our economic analysis.

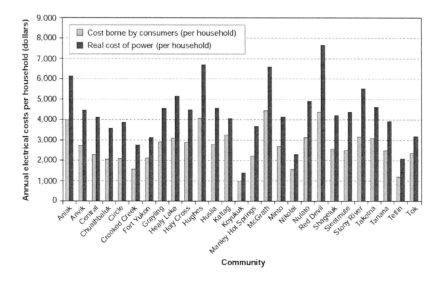

Figure 11. Annual village electrical costs, expressed on a per-household basis. These figures include costs incurred for electrical use in private homes as well as in shared facilities such as schools, tribal offices, and washeterias. The discrepancy between the cost borne by consumers and the real cost of power is covered by government funding, primarily via the Power Cost Equalization Program.

The gap between real and realized costs has negative social ramifications, creating disincentives for locally based efficiency improvements, sustainable community planning, and innovative use of capital (Colt et al. 2003). Even if biomass fuel use can be shown to be an option that is feasible in a given community, village residents may lack the necessary economic incentive to catalyze change. Moreover, the small population base of most villages has in the past proven to be an obstacle to reliably securing the necessary human resources for governance, operation, and maintenance of utilities (Colt et al. 2003). On the other hand, although government entities may have a financial incentive to promote change and may have the necessary technical expertise and human resources, they may suffer from bureaucratic inertia and lack of social impetus. Based on the financial power wielded at higher levels of governance and the social power contained within communities, there are potential advantages and disadvantages associated with both top-down or bottom-up approaches to managing potential village biomass projects (table 10)

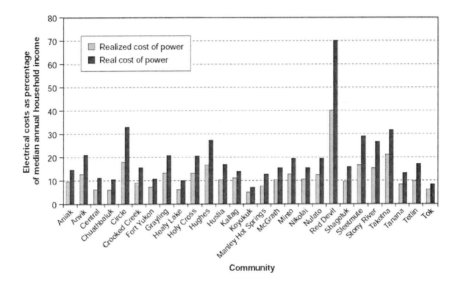

Figure 12. Village electrical costs per household expressed as a percentage of median household income for each community. Figures include costs incurred for electrical use in private homes as well as in shared facilities such as schools, tribal offices, and washeterias. The discrepancy between the costs borne by consumers ("realized costs") and the total unsubsidized costs ("real cost of power") is covered by government funding, primarily via the Power Cost Equalization Program.

At the state and federal levels, grants and other sources of funding are available to cover startup costs, and technical expertise is available for design and implementation work, including funds specifically allocated to renewable energy and alternative power (AEA 2005). Most of these funds would likely be channeled through the AEA, as detailed in table 7.

The advantages of the infrastructural assistance and funding available through AEA give rural Alaska a potential edge over rural communities in less developed nations, where capital and technological inputs are more uniformly scarce. Even in India, a nation with a stronger economy than many developing nations, lack of financial support for technology improvements has been cited as the primary reason for failure of an early attempt at instituting a small-scale biomass energy project (Kishore et al. 2004). A national-level analysis in India showed that biomass gasifiers 20 to 200 kW in capacity could entirely meet rural electricity needs (Somashekhar et al. 2000). Some demonstration projects have proven relatively successful (Somashekhar et al. 2000) whereas others have not (Kishore et al. 2004). There are several reasons for project failure, including subsidized power available from the existing grid, extremely low purchasing power among village residents, and poor technology for burning biomass other than wood chips (such as rice husks and other plant residues) (Kishore et al. 2004). Because Alaska's villages are largely removed from the power grid, have greater cash flows than rural Indian communities, and have wood as the primary source of potential biomass, these problems are unlikely to be applicable.

Table 10. Advantages and disadvantages of top-down vs. bottom-up strategies for implementing a fuel-conversion program

	Advantages	Disadvantages
Federal government	Power to limit carbon emission laws and treaties	Poor understanding of Alaska
State government	Power to create a statewide program	Emphasis on state rather than commmunity needs
Native corporations	Available capital; interest in village investments	Limited to for-profit activities; no statewide mission
Power cooperatives	Technical knowledge; statewide linkages	Commitment to existing diesel infrastructure
Village councils	Understanding of community needs	Lack of economic and human resources

In addition to providing funding and know-how, governments may be the most effective managers of some aspects of on-the-ground efforts. Some degree of centralization and top-down effort are predicated by the tiny size of some of the communities in question. For example, specialized skills such as boiler design and installation and engineering of combined heat and power grid systems would not be found in every community of 50 to 100 individuals.

However, direct management from the state or federal level is rife with potential problems. The same remoteness that makes the cost of diesel fuel in villages so high also demands that village power and heating systems be internally rather than externally managed whenever possible. Cultural considerations bolster this assertion. Village residents, most of them Alaska Natives, strongly prefer local control of village affairs (Hanson 2005, Putnam 2005).

Not only is local autonomy culturally preferable, it is also likely to be crucial for the long-term viability of biomass projects. State and federal officials are unlikely to be knowledgeable concerning important details such as interpersonal dynamics in the community, traditional use in the area around the village slated for harvest, and local concerns regarding fire risk. For example, during community studies preliminary to the installation of a biomass energy system in the village of McGrath, residents expressed concerns about the technical and economic feasibility of the project; the impacts of increased wood harvest on subsistence activities, aesthetics, and future wood supply; and overall system complexity (Crimp and Adamian 2001). Alaska Natives are often suspicious of solutions derived by govern-mental groups that are perceived to be part of the problem, and without community support, trust, and buy-in, programs instituted by outside entities are doomed to failure (Reiger et al. 2002).

In addition, local residents are likely to be able to provide realistic assessments of what type of employment would be considered desirable, and on what time scale it might be undertaken. For example, wood harvest, chipping, and transport might be shared informally among several individuals, and might be timed not only to coincide with adequate snowpack for easy transport, but also to fit in with seasonal subsistence activities and other seasonal employment. In most communities, gathering wood fuel is already part of subsistence activities; community members would be best equipped to decide how and when to expand fuel collection and how to pay individuals for the wood they gather. Since fuel gathering would be coupled with fire prevention, and because fuel collection would be most likely to occur in the winter via snowmachine rather than in the summer fire season, existing fire crews would be an obvious choice of labor force. Hanson (2005) noted that fire crews were involved in fuel clearing projects in Healy Lake, Tanacross, and Stevens Village, and that these groups generally work well together and are actively interested in fire protection. However, he also commented that work crews vary, and that having a good crew boss or leader is crucial to success.

Village councils, local light and power cooperatives, and Native corporations have greater power to implement projects than do individuals. For example, these entities are eligible for state or federal grants such as those being made available through the Alaska Wood Energy Development Task

Group. These grants, however, are being channeled via AEA. At an intermediate level of governance, organizations such as AEA, AVEC, and other regional light and power cooperatives have the potential to help link the resources of governmental agencies with the resources of communities. These organizations have already taken a lead in proposing, funding, and implementing alternative energy projects (AEA 2005, AVEC 2005). Thus far, AVEC has focused on wind and hydroelectric power, as many of its customer communities are coastal. Also, AEA has taken a lead in biomass demonstration projects, including installation of a wood-fired boiler in Dot Lake, and a proposal for a larger system in McGrath. The AEA has garnered funding for such projects from state and federal levels, but is implementing them using criteria that take into account local needs and local capacities.

In the long run, a combined approach seems likely to provide the greatest resilience to the system. Power sharing and co-management are ideas that are starting to take hold in a range of rural applications and are likely to be appropriate in an Alaskan context (Reiger et al. 2002). For example, although overarching assessments of fuel supply and demand around a village might be performed by forestry professionals, annual harvest areas might be chosen by local village councils, based on community preferences.

Based on the above information, we identified the following barriers and thresholds to change.

Barriers:

- The majority of AEA funding is traditionally allocated to existing system components, not to renewable energy or new technology startup.
- In some cases, state or AEA capital funds are designated for programs such as PCE and bulk fuel revolving loans, which create negative economic externalities favoring the status quo.
- Many power cooperatives are managed regionally, not at the village level.
- No forest certification system is in place whereby carbon credits could immediately be secured (although the potential for development of such a program exists within either the Alaska DNR or native corporation programs such as the TCC Forestry Program).
- Failure by the United States to sign onto binding climate-change agreements may keep carbon credit prices an order of magnitude lower here than overseas.

Thresholds:

- Existence of individuals within a given village who are willing and able to participate in fuel conversion projects, particularly village leaders who are willing to advocate for a biomass program, fire crews or other individuals actively interested in employment and fire prevention, and one or more crew leaders who can take responsibility for followthrough.
- Existence of necessary skills within village and the willingness of system operators to receive training in new technologies.
- Formation of effective cross-level collaboration, particularly between AEA (the likely funding agency and potential overarching project manager), village electrical companies or cooperatives (the likely applicants for funding and local managers), and individuals employed on the ground at the village level.

DISCUSSION

The transition to renewable energy sources is constrained by a number of economic, social, technological, and political factors. These include startup costs for research and new infrastructure; social inertia and risk aversion; inadequately developed technologies; lack of availability of all energy sources in all regions; and artificially low costs of existing fossil-fuel systems owing to subsidies, lack of accounting for economic externalities, and current infrastructure. Nevertheless, our results indicate that even with conservative assumptions for ecological, economic, and social parameters, conversion to wood biomass energy is likely to be a feasible and attractive option for many communities in interior Alaska. A successful fuel-conversion program must fulfill the social, economic, and ecological needs of the system as a whole (fig.13).

Based on our model, the communities likely to show the greatest ecological feasibility for biomass conversion are those in the small to medium size categories. Only the largest communities—those with populations over about 300—potentially lack adequate wood resources for complete fuel conversion within an easily accessible radius. This pattern runs counter to the trend whereby other services such as schools, clinics, and airports are more cost-effective in larger communities, leading to governmental pressure toward consolidation of small villages. Ironically, many villages have shrunk in part because of the high costs of fuel (DeMarban 2006b).

The greatest economic feasibility is demonstrated by villages with the highest benefit/cost ratio, which tend to be those not easily reached by either road or river networks. For these villages, even high estimates of costs for biomass fuel systems show an advantage over existing high costs for fuel transportation and storage.

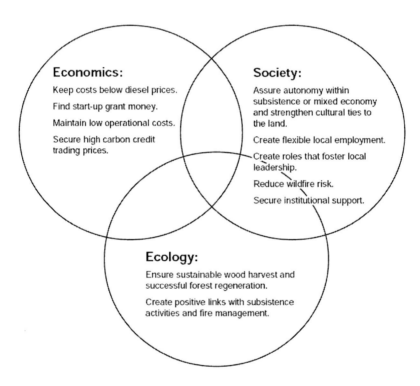

Figure 13. Social, economic, and ecological parameters affecting a potential fuel conversion program. These parameters are interconnected and subject to change over time.

Social feasibility, because it is so dependent on individuals, has yet to be determined on a village-by-village basis. However, it is likely to be greatest in communities with strong leadership, close ties to the land and its resources, and a core group of individuals—perhaps an existing fire crew—willing and able to work consistently on fuels harvest and associated tasks. These requirements tend to point toward medium or larger communities in remote areas.

Villages that fit both the ecological and the economic criteria include Alatna and Allakaket, Anvik, Central, Chuathbaluk, Circle, Crooked Creek, Grayling, Healy Lake, Holy Cross, Huslia, Kaltag, Lime Village, Manley Hot Springs, Minto, Nikolai, Nulato, Red Devil, Shageluk, Sleetmute, Stony River, and Takotna. In the smallest communities in this group, the presence or absence of strong leadership and willing workforce would be particularly critical in determining the success of conversion. For example, Takotna lists zero unemployed individuals from its 29 residents over the age of 16 (appendix). On the other hand, Aniak and Tanana show a positive benefit/cost ratio, but have populations above 300. Projects in these communities would have to be more cautious regarding wood supply, harvest area, and overall energy use, or might optimally be based on only partial conversion to wood fuel. Meanwhile Evansville and Bettles, Koyukuk, and Tetlin easily met ecological criteria but were on the borderline in the economic analysis. Fort Yukon, McGrath, Northway, and Tok all showed mixed results. These four communities are all either much larger than the mean, or located on a readily accessible transportation corridor, or both. Although biomass conversion projects may be feasible in these locations, additional factors would need to be taken into account, including the possibility of procuring wood from slightly farther afield (via road or river), and the effects of biomass conversion on the larger and more complex economies of these communities. Finally, inadequate data were available to fully assess potential feasibility for Beaver, Dot Lake, Galena, Hughes, and Ruby.

Our analysis was intentionally conservative, and may therefore have underestimated potential advantages of conversion to biomass fuels. For example, 110-year forest rotations are longer than would likely be considered by communities seeking fire protection and habitat revitalization. Our estimates for biomass per acre, forest cover, and carbon credit prices were relatively low, and our estimates for biomass system installation costs were relatively high. Perhaps the greatest undercounting of potential system benefits stems from the fact that, although we assumed that installed systems would provide both power and heat, we accounted for only the savings afforded by replacing the existing power supply. Although heating could in most cases only be provided for centrally located buildings, the savings afforded would likely be substantial in communities that already have infrastructure to support combined heat and power distribution, and worth assessing even in those that do not. Including heat as a resource increases estimates of biomass generator efficiency from approximately 28 to 68 percent (table 3). Even if less than half of this additional benefit stream could be effectively captured, it would increase the overall energy realized by more than 50 percent. An increase in system benefits of this magnitude would make almost all fuel conversion options economically attractive. Another potential source of error may stem from the fact that off-book expenses associated with current diesel systems were not considered, although they are likely to account for approximately 26 percent of total costs (Colt et al. 2003).

Finally, all estimates were made using 2004 fuel costs, which are substantially lower than more recent costs (DeMarban 2006a, 2006b). Fuel costs may continue to rise, and federal and state subsidies may shrink or disappear. The incentives for fuel conversion at the village level are highest when fuel prices are highest, but lower fuel costs might trigger the removal of state subsidies, as state revenues are almost entirely dependent on oil prices. These changes would make fuel conversion increasingly appealing— including, in many cases, conversion of 100 percent of generation capacity rather than partial conversion.

In addition to the sources of uncertainty explored in this analysis, other factors could affect the feasibility and desirability of biomass conversion programs. New transportation corridors might lower the costs of fuel transport in some areas. Additional local employment opportunities might drive up local wages, thus raising harvest costs or reducing the potential workforce. Payback on capital investments could be affected by inflation, deflation, or rapid changes in interest rates.

On the other hand, grant money such as that available through the Wood Energy Task Force, the Denali Commission, or AEA's Wood Energy Development Program could help jump-start projects, and might make infrastructure costs less of a concern. New technology might reduce the installation and operation costs for wood gasifiers below the range predicted, or international turmoil might cause fuel prices to skyrocket above predicted values. Carbon credit prices would eventually rise to match current ECX prices, even in the United States, if new binding international agreements are reached. Moreover, if fire on the landscape is perceived as an ever-increasing threat, and if state and federal firefighting resources become strained, then forest clearing might become more socially desirable and financially lucrative in its own right.

Many of these potential changes or surprises would tend to increase the economic viability of fuel conversion. However, model uncertainty not only means that economic outcomes are ambiguous for many villages, but also that social feasibility is uncertain. Thus, pilot projects offer the next step in testing feasibility. Such projects would help to validate our model, test technology under new conditions (e.g., remote location, cold climate), provide positive lessons that could be incorporated into future projects, and provide experience regarding errors to avoid.

When ecological, economic, and social parameters are considered in conjunction with one another, a pattern of hurdles and benefits emerges (table 11). Although many of these have been addressed in our analysis, others can only be truly tested through use of real-life project implementation.

Table 11. Potential hurdles and benefits associated with biomass fuel conversion in interior Alaska

	Hurdles	Benefits
Economic	Cost of new infrastructure Cost of biomass harvest	Wages from fuel gathering Reduced cost of diesel
Social/political	Political buy-in from agencies and power companies Ensuring local involvement and continuity	Health benefits from reduced pollution Greater autonomy of local communities
Technical/ecological	Technical challenges of biomass energy generation Ensuring long-term sustainability of harvest	Reduced fire risk Greater landscape diversity Creation of diverse wildlife habitat

Two existing pilot projects in interior Alaska demonstrate the feasibility of wood biomass systems and the efficacy of employing combined heat and power capabilities. The first, a wood-fired boiler used to heat and power eight residences and the washeteria in the 37-person community of Dot Lake, is already operational. The second, in McGrath, has not yet been completed but is slated to include a combined heat and power system based on continued use of diesel with a wood boiler providing additional energy to the system.

Dot Lake is not a typical interior village, as it is on the road system. As a result, diesel fuel in the community is far less expensive than in some villages, and our calculations show a strongly negative incentive for biomass fuels conversion. Nevertheless, Village Council President Bill Miller estimates that the village saves $6,500 to $13,000 in fuel costs per year using the wood-powered system (AEA 2000a). Capital costs were paid by external funding sources. However, wood prices in Dot Lake are not likely to be equivalent to prices in more remote villages, because in Dot Lake the boiler operates on wastes from nearby timber operations, which can be easily transported via road.

In McGrath, the option selected appeared economically preferable to three other possibilities: the status quo (all diesel); a wood boiler powering only the school; or a more comprehensive wood system, with diesel remaining as the backup fuel (Crimp and Adamian 2001). Crimp and Adamian (2001) also noted that the cost-effective use of biomass was highly dependent on the availability of inexpensive wood wastes; costs would be expected to rise sharply if roundwood harvest were required to operate the facility. However, at the time the analysis was done, it was assumed that the cost of bulk diesel would remain static at $1.54/gal ($0.41/L). In reality, prices have risen sharply, increasing by over 65 percent between 2003 and 2005, and potentially reaching $6/gal ($1.59/L) in 2008 (Bradner 2005, 2008).

As previously described, the potential income from sale of carbon credits from interior villages would be roughly $62,000 annually at 2005 market prices. In very small villages, the totals would be less than $300 per year. Even in larger communities, these sums represent only a very small percentage of the funds that would be necessary to operate and maintain combined heat and power systems of any kind. However, in some cases, these sums are enough to tip the balance toward biomass fuel conversion. If the value of carbon credits in the United States ever rises to meet world standards, perhaps because of future international agreements, the additive value of these credits could become a significant part of the cash economy at the village scale.

CONCLUSION

Given the combined drivers of rising fuel prices, ongoing climate change, increasing fire risk, and social pressures favoring fossil fuel independence, many communities may soon consider shifting to alternative fuels. The incentive of earning tradable carbon credits has added to potential benefit streams, and the monetary gains of participating in carbon markets may increase tenfold or more in the long term if the United States eventually implements programs congruent with those being used by Kyoto Protocol signatory nations.

In rural Alaska villages, economic conditions make fossil fuel use unusually expensive, and social conditions favor autonomy and local employment. Ecological conditions are likely to allow for harvesting a sustainable fuel source in a manner that enhances rather than detracts from ecological resilience, owing to the complex relationship between fire, forest succession, forest resources, fire suppression, and human settlements. Biomass fuels are likely to increase the long-term social and ecological resilience of village communities to externally-driven changes, including fluctuations in fossil fuel prices related to state, national, or international policies; variability in Alaska's economic outlook, which might in turn affect subsidies; and changes in fire risk and fire management, driven by climate change and by state and federal fire budgets.

For all of these reasons, interior Alaska village communities are in a position to be at the forefront in developing biomass fuels programs. Villages selected based on our combined ecological and social model would almost certainly reap benefits from the transition. In addition, because of the existence of substantial economic and political infrastructure at the state and federal levels, Alaska's rural communities are in a position to serve as pilot projects and leaders in a global movement toward rural biomass power.

ACKNOWLEDGMENT

The authors thank the following programs and individuals at the University of Alaska Fairbanks for their assistance and support:

- The Regional Resilience and Adaptation Program, part of the Integrative Graduate Education and Research Traineeship program funded by the National Science Foundation
- The Center for Global Change
- The Department of Biology and Wildlife
- Institute of Arctic Biology
- Janice Dawe, Jonathan Rosenberg, and A. David McGuire

ENGLISH EQUIVALENTS

When you know:	Multiply by:	To find:
Meters (m)	3.28	Feet

ENGLISH EQUIVALENTS (CONTINUED)

When you know:	Multiply by:	To find:
Kilometers (km)	0.621	Miles
Hectares (ha)	2.47	Acres
Liters (L)	0.265	Gallons (gal)
Kilograms (kg)	2.205	Pounds
Tonnes (t)	1.102	Tons
Tonnes per hectare (t/ha)	893	Pounds per acre
Square meters per hectare (m^2/ha)	4.37	Square feet per acre
Degrees Celsius (°C)	$(1.8 \times °C) + 32$	Degrees Fahrenheit
Kilograms per cubic meter (kg/m^3)	0.0624	Pounds per cubic foot
Joules (J)	0.000948	British thermal units (Btu)
Kilowatts (kW)	1.34	Horsepower
Kilowatt-hours (kWh)	3412	British thermal units
Kilowatt-hours per tonne (kWh/t)	0.645	British thermal units per pound

REFERENCES

Arctic Climate Impact Assessment [ACIA]. 2005. Arctic Climate Impact Assessment—Scientific Report. New York: Cambridge University Press. 1042 p.

Adamian, S.; Elliot, G.; Morris, G. [et al.]. 1998. The potential use of small biomass power technology to provide electricity for a Native Alaskan village. On file with: Serge Adamian, Ecotrade, Inc., 220 S Kenwood St., No. 305, Glendale, CA 91205.

Alaska Department of Community and Economic Development [ADCED]. 2005. Community Database Online. http://www.commerce.state.ak.us/dca/commdb/CF_BLOCK.htm. (November 3, 2005).

Alaska Energy Authority [AEA]. 2000a. McGrath Community Biomass Heating Project. Grant proposal to the U.S. Department of Energy. Anchorage, AK.

Alaska Energy Authority [AEA]. 2000b. Statistical report of the Power Cost Equalization Program. Anchorage, AK.

Alaska Energy Authority [AEA]. 2002. Statistical report of the Power Cost Equalization Program. Anchorage, AK.

Alaska Energy Authority [AEA]. 2004. Statistical report of the Power Cost Equalization Program. Anchorage, AK.

Alaska Energy Authority [AEA]. 2005. Alternative energy and energy efficiency assistance plan July 1, 2005 to June 30, 2007. Anchorage, AK.

Alaska Energy Authority [AEA]. 2007. Alternative energy and energy efficiency assistance plan July 1, 2007 to June 30, 2009. Anchorage, AK.

Alaska Department of Community and Economic Development. 2001. FY2002 Governor's operating budget—energy operations component. Juneau, AK.

Alaska Department of Community and Economic Development. 2002. State of Alaska FY2003 Governor's operating budget—energy operations component. Juneau, AK.

Alaska Village Electric Cooperative, Inc. [AVEC]. 2005. http://www.avec.org. (August 1, 2005).

Bain, R.L.; Overend, R.P.; Craig, K.R. 1996. Biomass-fired power generation. Snowbird, UT: Engineering Foundation; National Renewable Energy Laboratory.

Bain, R.L.; Amos, W.A.; Downing, M.; Perlack, R.L. 2003. Highlights of biopower technical assessment: state of the industry and the technology. Oak Ridge, TN: National Renewable Energy Laboratory, Oak Ridge National Laboratory.

Berkowitz, E. 2004. An act relating to carbon sequestration; and providing for an effective date. Alaska Statutes. AS.44.37.200-220.

Bradner, T. 2005. Worries over fuel costs may turn to nightmares. *Alaska Journal of Commerce*. July 24.

Bradner, T. 2008. Skyrocketing fuel costs drain rural Alaska bank accounts. Capital City Weekly. June 11.

Chicago Climate Exchange [CCX]. 2006. http://www.chicagoclimatex.com. (May 2, 2006).

Colt, S.; Goldsmith, S; Witta, A. 2003. Sustainable utilities in rural Alaska: effective management, maintenance, and operation of electric, water, sewer, bulk fuel, solid waste. Final Report. Anchorage, AK: Institute of

Social and Economic Research, University of Alaska Anchorage; Mark A. *Foster and Associates*. 36 p.

Crimp, P.M.; Adamian, S.V. 2001. Biomass energy alternatives for a remote Alaskan community. Anchorage, AK: Alaska Energy Authority; Glendale, CA: Ecotrade, Inc.

Crimp, P.M. 2005. Wood fired boilers for rural communities. http://www.uaf. edu/aetdl/woodfiredboilers.pdf. (September 3, 2005).

DeMarban, A.2006a. Out of the bush. *Anchorage Daily News.* February 28: A1.

DeMarban, A. 2006b. Powering down: the new Dark Ages. *Anchorage Daily News.* April 4: B1.

Demirbas, A. 2004. Current technologies for the thermo-conversion of biomass into fuels and chemicals. *Energy Sources.* 26(8): 715–30.

Department of the Interior, Bureau of Land Management [USDI BLM]. 2005. Snapshots: successful BLM projects supporting the National Fire Plan. Washington, DC. 6 p.

Devine, M.; Baring-Gould, E.I.; Petrie, B. 2005. Wind-diesel hybrid options for remote villages in Alaska. Anchorage, AK: National Renewable Energy Lab; Alaska Village Electric Cooperative; *Alaska Energy Authority.* 12 p.

DeWilde, L.; Chapin, F.S., III. 2006. Human impacts on the fire regime of interior Alaska: interactions among fuels, ignition sources, and fire suppression. *Ecosystems.* 9(8): 1342–1353.

Duval, J.E. 2004. Market opportunities for carbon sequestration in Alaska. Fairbanks AK: University of Alaska Fairbanks. M.S. thesis.

Dyrness, C.T., Viereck, L.A.; van Cleve, K. 1986. Fire in taiga communities of interior Alaska. In: Forest ecosystems in the Alaskan taiga. van Cleve, K.; Chapin, F.S., III; Flanagan, L.B.; Viereck, L.; Dyrness, C.T., eds. New York: Springer-Verlag: 74–86.

Energy Information Administration [EIA]. 2005. Electric Power Monthly. http://www.eia.doe.gov/cneaf/electricity/epm/epm_sum.html. (July 28, 2005).

Engineering News-Record [ENR]. 2001. Small generator is chip-fired. 246(20): 32.

Environmental Protection Agency [EPA]. 1999. Activities in Region 10, Alaska operations. Oil Spill Program Update: The U.S. EPA's Oil Program Center Report 3(1).

Fitzsimmons, M. 2003. Effects of deforestation and reforestation on landscape spatial structure in boreal Saskatchewan, Canada. *Forest Ecology and Management.* 174(1–3): 577–592.

Government Accounting Office [GAO]. 2005. Federal agencies are engaged in various efforts to promote the utilization of woody biomass, but significant obstacles to its use remain. Washington, DC.

Golden Valley Electric Association [GVEA]. 2005. http://www.gvea.com. (July 28 2005.)

Hansen, J.; Nazarenko, L.; Ruedy, R. [et al.] 2005. Earth's energy imbalance: confirmation and implications. *Science.* 308(5727): 1431–1435.

Hanson, D. 2005. Personal communication. Forest resource manager, Alaska Department of Natural Resources Division of Forestry, 3700 Airport Way, Fairbanks, AK 99709.

Haq, Z. 2002. Evaluating biomass for electricity generation. *BioCycle.* 43(11): 33–6.

Hollingsworth, T. N. 2004. Quantifying variability in the Alaskan black spruce ecosystem: linking vegetation, carbon, and fire history. DAI, 66, no. 01B. Fairbanks, AK: University of Alaska Fairbanks.

Houghton, J.T.; Ding, Y.; Griggs, D.J. [et al.], eds. 2001. Climate change 2001: the scientific basis. Third assessment report of the Inter-governmental Panel on Climate Change. Contribution of Working Group I. Cambridge, UK: Cambridge University Press. 881 p.

Innes, J.; Peterson, D. 2001. Managing forests in a greenhouse world—context and challenges. Climate change, carbon, and forestry in northwestern North America. Orcas Island, WA: U.S. Department of Agriculture.

Karl, T.R.; Trenberth, K.E. 2003. Modern global climate change. *Science.* 302(5651): 1719–1723.

Kirk, L. 2004. Russia considers setting up EU-style emissions trading scheme. *EU Observer.* November 4.

Kishore, V.V.N.; Bhandari, P.M.; Gupta, P. 2004. Biomass energy technologies for rural infrastructure and village power—opportunities and challenges in the context of global climate change concerns. *Energy Policy.* 32(6): 801–810.

Lee, M. 2005. Reducing hazardous fuels. Under the Canopy: Forestry and Forest Products Newsletter of the Cooperative Extension Service. March: 15–16.

Mark A. Foster and Associates [MAFA]; Northern Economics, Inc. 2004. Alaska Rural Energy Plan; initiatives for improving energy efficiency and reliability. Anchorage, AK: *Alaska Energy Authority.* 24 p.

Maker, T.M. 2004. Wood-chip heating systems: a guide for institutional and commercial biomass installations. Montpelier, VT: Biomass Energy Resource Center. 91 p.

McCrone, A. 2005. Carbon trading takes off. The Australian. August 17. Science and Nature section.

McIlveen-Wright, D.R.; McMullan, J.T.; Gainey, D.J. 2003. Wood-fired fuel cells in selected buildings. *Journal of Power Sources.* 118(1–2): 393–404.

McNamara, W. 2004. CO2 emissions trading: nascent market grows despite regulatory uncertainty. *Power Engineering.* 108(8): 52–56.

Omosun, A.O.; Bauen, A.; Brandon, N.P. 2004. Modelling system efficiencies and costs of two biomass-fuelled SOFC systems. *Journal of Power Sources.* 131(1/2): 96–106.

Poe, R.G. 2001. State of Alaska FY2003 Governor's operating budget. Juneau, AK: Budget Request Unit, Department of Community and Economic Development, Alaska Energy Authority.

Poe, R.J. 2002. Bulk fuel systems upgrades. State of Alaska appropriation request, Health/Safety. http://www.gov.state.ak.us/omb/2002site/Budget/ DCED/proj32584.pdf. (January 11, 2006.)

Prentice, I.C.; Heimann, M.; Sitch, S. 2000. The carbon balance of the terrestrial biosphere: ecosystem models and atmospheric observations. *Ecological Applications.* 10(6): 1553–1573.

Prestemon, D.R. 1998. Determining moisture content of wood. http://www. ag.iastate.edu/departments/forestry/ext/pubs/F-352.pdf.   (January   11, 2006.)

Putnam, W. 2005. Hazard fuel treatment projects. Personal communication. Forester, Tanana Chiefs Conference, 122 First Avenue, Ste. 600, Fairbanks, AK 99701.

Rees, D.C.; Juday, G.P. 2002. Plant species diversity on logged versus burned sites in central Alaska. *Forest Ecology and Management.* 155(1–3): 291–302.

Rieger, L.; Wood, D.S.; Jennings, M. 2002. Alaska Native Technical Assistance and Resource Center. Final Report. Anchorage, AK: Justice Center, University of Alaska Anchorage. 50 p.

Schwarzenegger, A. 2005. Executive Order S-3-05 by the Governor of the State of California.

Sharratt, B.S. 1997. Thermal conductivity and water retention of a black spruce forest floor. *Soil Science.* 162(8): 576–582.

Somashekhar, H.I.; Dasappa, S.; Ravindranath, N.H. 2000. Rural bioenergy centres based on biomass gasifiers for decentralized power generation:

case study of two villages in southern India. *Energy for Sustainable Development.* 4(3): 55–63.

University of Alaska Anchorage [UAA]. 2003. Alaska electric power statistics (with Alaska energy balance) 1960–2001. Anchorage, AK: Institute of Social and Economic Research, University of Alaska Anchorage.

University of Alaska Fairbanks [UAF] 2005. Alaska Energy News. http://www. uaf.edu/energyin/webpage/pages/alaska%20energy%20news.htm. (November 11, 2005).

United Nations [UN]. 1997. Kyoto protocol to the United Nations framework on climate change, Conference of the Parties, Third Session. Kyoto, Japan.

U.S. Department of Agriculture, Forest Service [USDA FS]. 2004. Wood biomass for energy. Techline. Madison, WI: Forest Products Laboratory, State and Private Forestry Technology Marketing Unit. 3 p.

Waldheim, L.; Carpentieri, E. 2001. Update on the progress of the Brazilian wood BIG-GT demonstration project. *Journal of Engineering for Gas Turbines and Power.* 123(3): 525–536.

Willeboer, W. 1998. The Amer demolition wood gasification project. *Biomass and Bioenergy.* 15(3): 245–249.

Wu, Z.; Wu, C.; Huang, H. 2003. Test results and operation performance analysis of a 1-MW biomass gasification electric power generation system. *Energy and Fuels.* 17(3): 619–24.

Yarie, J.; Billings, S. 2002. Carbon balance of the taiga forest within Alaska: present and future. *Canadian Journal of Forest Research—Revue Canadienne De Recherche Forestiere.* 32(5): 757–767.

Yarie, J.; Mead, D. 1982. Aboveground tree biomass on productive forest land in Alaska. Res. Pap. PNW-298. Portland, OR. U.S. Department of Agriculture, Forest Service, Pacific Northwest Forest and Range Experiment Station. 16 p.

Zerbin, W.O. 1984. Generating electricity by gasification of biomass. Thermochemical Processing of Biomass: 297–306.

APPENDIX

Alaska Communities

Community	Access	Population[a]	Electric utility[a]	Total households[a]	Average number in household	Median household income[a]	Population 16 and over
						Dollars	
Alatna and Allakaket	Koyukuk	122	Alaska Power Company	53	2.30	n/a	89
Aniak	Kuskokwim	532	Aniak Light & Power Company	174	3.29	41,875	398
Anvik	Yukon	101	AVEC	39	2.67	21,250	69
Beaver	Yukon	67	Beaver Joint Utilities	31	2.71	28,750	86
Evansville and Bettles	Koyukuk	51	Alaska Power Company	28	1.82	n/a	66
Central	Minor Road	102	Central Electric, Inc.	67	2.00	36,875	113
Chuathbaluk	Kuskokwim	105	Middle Kuskokwim Electric Cooperative	33	3.61	34,286	90
Circle	Minor Road	99	Circle Electric Utility	34	2.94	11,667	50
Crooked Creek	Kuskokwim	147	Middle Kuskokwim Electric Cooperative	38	3.61	17,500	90
Dot Lake	Major Road	29	Alaska Power Company	10	1.90	13,750	18
Eagle and Eagle Village	Minor Road	183	Alaska Power Company	90	2.03	n/a	140

Community	Access	Population[a]	Electric utility[a]	Total households[a]	Average number in household[a]	Median household income[a] (Dollars)	Population 16 and over
Fort Yukon	Yukon	594	Gwitchyaa Zhee Utilities	225	2.62	29,375	449
Galena	Yukon	717	City of Galena	216	2.83	61,125	495
Grayling	Yukon	182	AVEC	51	3.80	21,875	105
Healy Lake	Minor River	34	Alaska Power Company	13	2.85	51,250	43
Holy Cross	Yukon	206	AVEC	64	3.55	21,875	165
Hughes	Koyukuk	72	Hughes Power & Light	26	3.00	24,375	50
Huslia	Koyukuk	269	AVEC	88	3.33	27,000	188
Kaltag	Yukon	211	AVEC	69	3.33	29,167	159
Koyukuk	Yukon	109	City of Koyukuk	39	2.59	19,375	68
Lime Village	Minor river	34	Lime Village Power	19	1.79	n/a	n/a
Manley Hot Springs	Road	73	Manley Utility Company,	36	2.00	29,000	60
McGrath	Kuskokwim	367	McGrath Light & Power	145	2.77	43,056	286
Minto	Minor Road	207	AVEC	74	3.49	21,250	179
Nikolai	Minor River	121	Nikolai Light & Power	40	2.50	15,000	60
Northway and Northway Village	Major Road	195	Alaska Power Company	62	3.15	n/a	159
Nulato	Yukon	320	AVEC	91	3.69	25,114	213
Red Devil	Kuskokwim	35	Middle Kuskokwim Electric Cooperative	17	2.82	10,938	29

Alaska Communities (continued)

Community	Access	Population[a]	Electric utility[a]	Total households[a]	Average number in household	Median household income[a]	Population 16 and over
						Dollars	
Ruby	Yukon	190	City of Ruby	68	2.76	24,375	119
Shageluk	Minor River	132	AVEC	36	3.58	26,667	76
Sleetmute	Kuskokwim	78	Middle Kuskokwim Electric Cooperative	33	3.03	15,000	52
Stony River	Kuskokwim	54	Middle Kuskokwim Electric Cooperative	19	3.21	20,714	49
Takotna	Kuskokwim	47	Takotna Community Association Utilities	19	2.63	14,583	29
Tanana	Yukon	304	Tanana Power Company	121	2.55	29,750	210
Tetlin	Minor Road	129	Alaska Power Company	42	2.79	12,250	70
Tok	Major Road	1,439	Alaska Power Company	534	2.61	37,941	995

n/a = not available, AVEC = Alaska Village Electrical Cooperative. *a* Data from ADCED 2005.

Alaska community fuel use and power generation

Community	Unemployed[a]	Fuel use (FY2004)[b]	Average price of diesel fuel (2004)[b]	Fuel costs	Installed capacity[c]	Power generated (2004)[b]	Average load	Average load/ installed capacity
		Gallons	Dollars/gallon	Dollars	kW	kWh	kW	
Alatna and Allakaket	20	53,773	2.19	117,763	430	648,861	74	0.17
Aniak	35	192,576	1.32	254,200	2,865	2,468,700	282	0.10
Anvik	11	38,474	1.32	50,786	337	469,023	54	0.16
Beaver	12	31,436	1.92	60,357	137	293,400	33	0.24
Evansville and Bettles	n/a	58,368	1.41	82,299	650	703,820	80	0.12
Central	8	50,104	1.22	61,127	640	501,896	57	0.09
Chuathbaluk	3	20,200	1.70	34,340	n/a	213,737	24	n/a
Circle	6	34,750	1.24	43,090	200	372,000	42	0.21
Crooked Creek	21	25,258	1.69	42,686	n/a	254,434	29	n/a
Dot Lake	2	n/a	n/a	n/a	325	n/a	n/a	n/a
Eagle and Eagle Village	25	58,474	1.20	70,169	477	781,344	89	0.19
Fort Yukon	52	207,698	1.66	344,779	2,400	2,840,000	324	0.14
Galena	32	724,076	1.46	1,057,151	6,000	9,466,799	1,081	0.18
Grayling	13	46,352	1.52	70,455	546	588,761	67	0.12
Healy Lake	5	14,339	1.25	17,924	105	152,986	17	0.17
Holy Cross	22	54,340	1.51	82,053	585	708,012	81	0.14

Alaska community fuel use and power generation (continued)

Community	Unemployed[a]	Fuel use (FY2004)[b]	Average price of diesel fuel (2004)[b]	Fuel costs	Installed capacity[c]	Power generated (2004)[b]	Average load	Average load/ installed capacity
		Gallons	Dollars/gallon	Dollars	kW	kWh	kW	
Hughes	3	37,325	3.27	122,053	323	n/a	n/a	n/a
Huslia	21	77,648	1.79	138,990	680	916,941	105	0.15
Kaltag	29	57,498	1.58	90,847	573	663,172	76	0.13
Koyukuk	12	20,830	1.89	39,369	244	353,250	40	0.17
Lime Village	n/a	9,101	4.44	40,408	77	99,263	11	0.15
Manley Hot Springs	4	26,772	1.14	30,520	480	294,120	34	0.07
McGrath	24	221,650	1.40	310,310	2,685	2,963,200	338	0.13
Minto	29	56,366	1.13	63,694	558	722,562	82	0.15
Nikolai	11	38,182	1.81	69,109	362	401,400	46	0.13
Northway and Northway Villlage	19	121,569	1.29	156,824	1,165	1,583,944	181	0.16
Nulato	52	85,982	1.59	136,711	897	1,148,831	131	0.15
Red Devil	4	14,490	1.83	26,517	173	126,434	14	0.08
Ruby	17	24,861	1.76	43,755	654	n/a	n/a	n/a
Shageluk	17	31,506	1.69	53,245	370	405,639	46	0.13
Sleetmute	8	25,314	1.69	42,781	208	229,258	26	0.13
Stony River	8	13,994	1.69	23,650	139	116,418	13	0.10

Community	Unemployed[a]	Fuel use (FY2004)[b]	Average price of diesel fuel (2004)[b]	Fuel costs	Installed capacity[c]	Power generated (2004)[b]	Average load	Average load/installed capacity
		Gallons	Dollars/gallon	Dollars	kW	kWh	kW	
Takotna	0	28,219	1.72	48,537	297	248,705	28	0.10
Tanana	31	104,270	1.34	139,722	1,456	1,378,060	157	0.11
Tetlin	15	40,782	1.46	59,542	280	473,310	54	0.19
Tok	111	861,311	1.25	1,076,639	4,960	12,518,973	1,429	0.29

n/a = not available.

[a] Data from ADCED 2005.
[b] Data from AEA 2004; nonfuel expenses for Alaska Village Electrical Cooperative (AVEC) villages are calculated at the average rate for the cooperative.
[c] Data from UAA 2003.

Alaska community power costs

Community	Power per capita	Total nonfuel expenses (2004)[a]	PCE payments (2004)[a]	Residential rate without PCE[a]	Residential rate after subsidy[a]	Real cost of power	Real cost
	kWh	Dollars	Dollars	Dollars per kWh	Dollars per kWh	Dollars	Dollars per kWh
Alatna and Allakaket	5319	83,371	84,787	0.48	0.27	201,134	0.31
Aniak	4640	735,336	168,391	0.49	0.32	989,536	0.40
Anvik	4644	117,256	47,007	0.46	0.28	168,041	0.36
Beaver	4379	n/a	17,620	0.42	0.26	n/a	n/a

Alaska community power costs (continued)

Community	Power per capita	Total nonfuel expenses (2004)[a]	PCE payments (2004)[a]	Residential rate without PCE[a]	Residential rate after subsidy[a]	Real cost of power	Real cost
	kWh	Dollars	Dollars	Dollars per kWh	Dollars per kWh	Dollars	Dollars per kWh
Evansville and Bettles	13 800	74,967	34,316	0.41	0.20	157,266	0.22
Central	4921	148,543	63,922	0.51	0.28	209,670	0.42
Chuathbaluk	2036	69,482	37,319	0.56	0.32	103,822	0.49
Circle	3758	86,608	37,593	0.50	0.27	129,698	0.35
Crooked Creek	1731	68,424	44,743	0.56	0.32	111,110	0.44
Dot Lake	n/a	15,551	9,751	0.23	0.17	n/a	n/a
Eagle and Eagle Village	4270	128,692	65,932	0.41	0.26	198,861	0.25
Fort Yukon	4781	362,638	142,391	0.34	0.23	707,417	0.25
Galena	13 203	n/a	124,170	0.25	0.18	n/a	n/a
Grayling	3235	147,190	69,919	0.44	0.28	217,645	0.37
Healy Lake	4500	43,540	13,490	0.40	0.24	61,464	0.40
Holy Cross	3437	177,003	83,911	0.42	0.27	259,056	0.37
Hughes	n/a	38,238	27,077	0.51	0.30	160,291	n/a
Huslia	3409	229,235	105,966	0.46	0.28	368,225	0.40
Kaltag	3143	165,793	70,921	0.46	0.28	256,640	0.39
Koyukuk	3241	18,747	12,804	0.45	0.36	58,116	0.16
Lime Village	2920	62,517	11,556	0.80	0.56	102,925	1.04

Community	Power per capita	Total nonfuel expenses (2004)[a]	PCE payments (2004)[a]	Residential rate without PCE[a]	Residential rate after subsidy[a]	Real cost of power	Real cost
	kWh	Dollars	Dollars	Dollars per kWh	Dollars per kWh	Dollars	Dollars per kWh
Manley Hot Springs	4029	103,826	34,735	0.60	0.36	134,346	0.46
McGrath	8074	561,359	162,757	0.43	0.29	871,669	0.29
Minto	3491	180,641	77,094	0.40	0.26	244,334	0.34
Nikolai	3317	42,004	47,474	0.50	0.34	111,113	0.28
Northway and Northway Village	8123	88,293	85,818	0.43	0.25	245,117	0.15
Nulato	3590	287,208	138,928	0.44	0.28	423,919	0.37
Red Devil	3612	68,461	16,839	0.56	0.32	94,978	0.75
Ruby	n/a	15,999	19,635	0.46	0.33	59,754	n/a
Shageluk	3073	101,410	42,971	0.46	0.28	154,655	0.38
Sleetmute	2939	69,424	41,057	0.56	0.32	112,205	0.49
Stony River	2156	69,067	16,594	0.56	0.32	92,717	0.80
Takotna	5292	33,897	20,849	0.48	0.32	82,434	0.33
Tanana	4533	326,127	109,284	0.49	0.31	465,849	0.34
Tetlin	3669	36,882	48,354	0.47	0.27	96,424	0.20
Tok	8700	671,543	212,194	0.23	0.17	1,748,182	0.14

Note: PCE = Power Cost Equalization program; n/a = not available.
[a] Data from AEA 2004; nonfuel expenses for Alaska Village Electrical Cooperative (AVEC) villages are calculated at the average rate for the cooperative.

Alaska community power cost per household and biomass system cost

	Real cost of power per household	Real cost of power per household as percentage of median household income	Estimated installed cost of biomass system ($1,849/kw)		
			To meet 50 percent of mean load	To meet mean load	To replace 100 percent of existing generation capacity
	Dollars	Percent	Dollars	Dollars	
Alatna and Allakaket	n/a	n/a	68,479	136,957	795,070
Aniak	6,120	14.6	260,538	521,076	5,297,385
Anvik	4,442	20.9	49,499	98,998	623,113
Beaver	n/a	n/a	30,964	61,929	253,313
Evansville and Bettles	n/a	n/a	74,279	148,557	1,201,850
Central	4,111	11.1	52,968	105,937	1,183,360
Chuathbaluk	3,569	10.4	22,557	45,114	n/a
Circle	3,852	33.0	39,260	78,519	369,800
Crooked Creek	2,729	15.6	26,852	53,704	n/a
Dot Lake	n/a	n/a	n/a	n/a	600,925
Eagle and Eagle Village	n/a	n/a	82,460	164,921	881,973
Fort Yukon	3,120	10.6	299,724	599,447	4,437,600
Galena	n/a	n/a	999,093	1,998,186	11,094,000
Grayling	4,544	20.8	62,136	124,272	1,009,554
Healy Lake	5,152	10.1	16,146	32,291	194,145
Holy Cross	4,464	20.4	74,721	149,442	1,081,665
Hughes	6,679	27.4	n/a	n/a	597,227

	Real cost of power per household	Real cost of power per household as percentage of median household income	Estimated installed cost of biomass system ($1,849/kw)		
			To meet 50 percent of mean load	To meet mean load	To replace 100 percent of existing generation capacity
	Dollars	*Percent*		*Dollars*	
Huslia	4,558	16.9	96,771	193,542	1,257.320
Kaltag	4,050	13.9	69,989	139,978	1,059,477
Koyukuk	1,381	7.1	37,281	74,562	451,156
Lime Village	n/a	n/a	10,476	20,952	142,373
Manley Hot Springs	3,681	12.7	31,040	62,081	887,520
McGrath	6,579	15.3	312,726	625,452	4,964,565
Minto	4,119	19.4	76,257	152,513	1,031.742
Nikolai	2,296	15.3	42,362	84,725	669,338
Northway and Northway Village	n/a	n/a	167,164	334,328	2,154,085
Nulato	4,888	19.5	121,244	242,487	1,658,553
Red Devil	7,652	70.0	13,343	26,687	319,877
Ruby	868	3.6	n/a	n/a	1,209.246
Shageluk	4,194	15.7	42,810	85,619	684,130
Sleetmute	4,359	29.1	24,195	48,390	384,592
Stony River	5,512	26.6	12,286	24,573	257,011
Takotna	4,613	31.6	26,247	52,495	549,153
Tanana	3,908	13.1	145,436	290,871	2,692,144

Alaska community power cost per household and biomass system cost (continued)

	Real cost of power per household	Real cost of power per household as percentage of median household income	Estimated installed cost of biomass system ($1,849/kw)		
			To meet 50 percent of mean load	To meet mean load	To replace 100 percent of existing generation capacity
	Dollars	Percent	Dollars		
Tetlin	2,085	17.0	49,951	99,903	517,720
Tok	3,171	8.4	1,321,209	2,642,418	9,171,040

n/a = not available.

Alaska community power system annual costs

Community	Annual operating cost of biomass system ($0.17/kWh)		Annual diesel fuel cost offset		Annual nonfuel cost offset		Estimated annual savings (compared to real cost of diesel system), mean load	Per capita annual savings, mean load
	50 percent of mean load	Mean load	50 percent of mean load	Mean load	50 percent of mean load	Mean load		
				Dollars				
Alatna and Allakaket	44,123	66,184	47,105	70,658	8,337	20,843	25,317	208
Aniak	167,872	251,807	101,680	152,520	73,534	183,834	84,547	159
Anvik	31,894	47,840	20,314	30,471	11,726	29,314	11,945	118

Community	Annual operating cost of biomass system ($0.17/kWh)		Annual diesel fuel cost offset		Annual nonfuel cost offset		Estimated annual savings (compared to real cost of diesel system), mean load	Per capita annual savings, mean load
	50 percent of mean load	Mean load	50 percent of mean load	Mean load	50 percent of mean load	Mean load		
			Dollars					
Beaver	19,951	29,927	24,143	36,214	n/a	n/a	n/a	n/a
Evansville and Bettles	47,860	71,790	32,920	49,379	7,497	18,742	-3,669	-72
Central	34,129	51,193	24,451	36,676	14,854	37,136	22,618	222
Chuathbaluk	14,534	21,801	13,736	20,604	6,948	17,371	16,173	154
Circle	25,296	37,944	17,236	25,854	8,661	21,652	9,562	97
Crooked Creek	17,302	25,952	17,074	25,612	6,842	17,106	16,765	114
Dot Lake	n/a	n/a	n/a	n/a	1,555	3,888	n/a	n/a
Eagle and Eagle Village	53,131	79,697	28,068	42,101	12,869	32,173	-5,423	-30
Fort Yukon	193,120	289,680	137,911	206,867	36,264	90,660	7,847	13
Galena	643,742	965,613	422,860	634,291	n/a	n/a	n/a	n/a
Grayling	40,036	60,054	28,182	42,273	14,719	36,798	19,017	104
Healy Lake	10,403	15,605	7,170	10,754	4,354	10,885	6,035	177
Holy Cross	48,145	72,217	32,821	49,232	17,700	44,251	21,266	103
Hughes	n/a	n/a	n/a	n/a	3,824	9,560	n/a	n/a
Huslia	62,352	93,528	55,596	83,394	22,924	57,309	47,175	175
Kaltag	45,096	67,644	36,339	54,508	16,579	41,448	28,313	134

Alaska community power system annual costs (continued)

Community	Annual operating cost of biomass system ($0.17/kWh)		Annual diesel fuel cost offset		Annual nonfuel cost offset		Estimated annual savings (compared to real cost of diesel system), mean load	Per capita annual savings, mean load
	50 percent of mean load	Mean load	50 percent of mean load	Mean load	50 percent of mean load	Mean load		
			Dollars					
Koyukuk	24,021	36,032	15,747	23,621	1,875	4,687	-7,724	-71
Lime Village	6,750	10,125	16,163	24,245	6,252	15,629	29,749	875
Manley Hot Springs	20,000	30,000	12,208	18,312	10,383	25,957	14,268	195
McGrath	201,498	302,246	124,124	186,186	56,136	140,340	24,279	66
Minto	49,134	73,701	25,477	38,216	18,064	45,160	9,675	47
Nikolai	27,295	40,943	27,644	41,466	4,200	10,501	11,024	91
Northway and Northway Village	107,708	161,562	62,730	94,094	8,829	22,073	-45,395	-233
Nulato	78,121	117,181	54,685	82,027	28,721	71,802	36,648	115
Red Devil	8,598	12,896	10,607	15,910	6,846	17,115	20,129	575
Ruby	n/a	n/a	n/a	n/a	1,600	4,000	n/a	n/a
Shageluk	27,583	41,375	21,298	31,947	10,141	25,352	15,924	121
Sleetmute	15,590	23,384	17,112	25,668	6,942	17,356	19,640	252
Stony River	7,916	11,875	9,460	14,190	6,907	17,267	19,582	363
Takotna	16,912	25,368	19,415	29,122	3,390	8,474	12,228	260

Community	Annual operating cost of biomass system ($0.17/kWh)		Annual diesel fuel cost offset		Annual nonfuel cost offset		Estimated annual savings (compared to real cost of diesel system), mean load	Per capita annual savings, mean load
	50 percent of mean load	Mean load	50 percent of mean load	Mean load	50 percent of mean load	Mean load		
				Dollars				
Tanana	93,708	140,562	55,889	83,833	32,613	81,532	24,803	82
Tetlin	32,185	48,278	23,817	35,725	3,688	9,221	-3,332	-26
Tok	851,290	1,276,935	430,656	645,983	67,154	167,886	-463,066	-322

n/a = not available.

Alaska community years to break even on capital investment

Community	Years to pay back capital (mean load, no carbon credits)	Years to pay back capital (mean load, CCX)	Years to pay back capital (mean load, ECX)	Estimated annual savings compared to real costs of diesel system, 1/2 mean load	Years to pay back capital (1/2 mean load, no carbon credits)	Years to pay back capital (1/2 mean load, CCX)	Years to pay back capital (1/2 mean load, ECX)	Potential annual	
								CCX prices	ECX prices
	Years			*Dollars*	*Years*			*Dollars*	
Alatna and Allakaket	5.4	5.3	4.3	11,320	6.0	5.9	4.4	860	10,862
Aniak	6.2	6.0	4.8	7,342	35.5	30.4	11.4	3,081	38,900
Anvik	8.3	8.0	6.0	146	338.4	126.1	15.2	616	7,772

Alaska community years to break even on capital investment (continued)

Community	Years to pay back capital (mean load, no carbon credits)	Years to pay back capital (mean load, CCX)	Years to pay back capital (mean load, ECX)	Estimated annual savings compared to real costs of diesel system, 1/2 mean load	Years to pay back capital (1/2 mean load, no carbon credits)	Years to pay back capital (1/2 mean load, CCX)	Years to pay back capital (1/2 mean load, ECX)	Potential annual CCX prices	Potential annual ECX prices
	Years	*Years*		*Dollars*		*Years*		*Dollars*	
Beaver	n/a	n/a	n/a	n/a	n/a	n/a	n/a	503	6,350
Evansville and Bettles	-40.5	-47.8	43.6	-7,444	-10.0	-10.5	-27.2	934	11,790
Central	4.7	4.6	3.7	5,176	10.2	9.6	5.7	802	10,121
Chuathbaluk	2.8	2.8	2.4	6,150	3.7	3.6	2.9	323	4,080
Circle	8.2	7.9	5.7	601	65.3	47.7	11.5	556	7,020
Crooked Creek	3.2	3.2	2.7	6,615	4.1	4.0	3.1	404	5,102
Dot Lake	n/a	n/a	n/a	n/a	n/a	n/a	n/a	n/a	n/a
Eagle and Eagle	-30.4	-33.9	99.1	-12,195	-6.8	-7.0	-11.0	936	11,812
Fort Yukon	76.4	60.9	18.2	-18,945	-15.8	-17.0	-138.6	3,323	41,955
Galena	n/a	n/a	n/a	n/a	n/a	n/a	n/a	11,585	146,263
Grayling	6.5	6.4	5.0	2,865	21.7	19.7	9.4	742	9,363
Healy Lake	5.4	5.2	4.2	1,120	14.4	13.3	7.1	229	2,896
Holy Cross	7.0	6.9	5.4	2,377	31.4	27.4	11.0	869	10,977

Community	Years to pay back capital (mean load, no carbon credits)	Years to pay back capital (mean load, CCX)	Years to pay back capital (mean load, ECX)	Estimated annual savings compared to real costs of diesel system, 1/2 mean load	Years to pay back capital (1/2 mean load, no carbon credits)	Years to pay back capital (1/2 mean load, CCX)	Years to pay back capital (1/2 mean load, ECX)	Potential annual CCX prices	ECX prices
	Years			*Dollars*		*Years*		*Dollars*	
Hughes	n/a	n/a	n/a	n/a	n/a	n/a	n/a	597	7,540
Huslia	4.1	4.0	3.4	16,168	6.0	5.8	4.3	1,242	15,685
Kaltag	4.9	4.8	4.0	7,822	8.9	8.5	5.6	920	11,615
Koyukuk	-9.7	-9.9	-14.3	-6,399	-5.8	-6.0	-7.9	333	4,208
Lime Village	0.7	0.7	0.7	15,665	0.7	0.7	0.6	146	1,838
Manley Hot Springs	4.4	4.3	3.5	2,590	12.0	11.2	6.5	428	5,408
McGrath	25.8	23.7	12.2	-21,238	-14.7	-15.8	-94.0	3,546	44,773
Minto	15.8	14.9	9.2	-5,593	-13.6	-14.6	-73.4	902	11,386
Nikolai	7.7	7.4	5.4	4,549	9.3	8.8	5.5	611	7,713
Northway and Northway Village	-7.4	-7.6	-10.9	-36,149	-4.6	-4.7	-6.3	1,945	24,557
Nulato	6.6	6.5	5.2	5,285	22.9	20.8	9.9	1,376	17,368
Red Devil	1.3	1.3	1.2	8,855	1.5	1.5	1.3	232	2,927
Ruby	n/a	n/a	n/a	n/a	n/a	n/a	n/a	398	5,022
Shageluk	5.4	5.3	4.3	3,856	11.1	10.6	6.7	504	6,364
Sleetmute	2.5	2.4	2.1	8,465	2.9	2.8	2.3	405	5,113

Alaska community years to break even on capital investment (continued)

Community	Years to pay back capital (mean load, no carbon credits)	Years to pay back capital (mean load, CCX)	Years to pay back capital (mean load, ECX)	Estimated annual savings compared to real costs of diesel system, 1/2 mean load	Years to pay back capital (1/2 mean load, no carbon credits)	Years to pay back capital (1/2 mean load, CCX)	Years to pay back capital (1/2 mean load, ECX)	Potential annual CCX prices	ECX prices
	Years			*Dollars*	*Years*			*Dollars*	
Stony River	1.3	1.2	1.2	8,450	1.5	1.4	1.3	224	2,827
Takotna	4.3	4.2	3.4	5,892	4.5	4.3	3.2	452	5,700
Tanana	11.7	11.3	7.8	-5,207	-27.9	-32.0	45.2	1,668	21,063
Tetlin	-30.0	-34.0	62.0	-4,680	-10.7	-11.3	-36.1	653	8,238
Tok	-5.7	-5.8	-7.4	-353,480	-3.7	-3.8	-4.7	13,781	173,985

Note: CCX = Chicago Climate Exchange, ECX = European Climate Exchange, n/a = not available. Negative years for payback indicate that payback will never occur; in such cases the transition to biomass fuels would not be profitable.

In: The Potential for Wood Energy ...   ISBN: 978-1-61470-990-9
Editor: Dmitry S. Halinen   © 2012 Nova Science Publishers, Inc.

Chapter 2

DEVELOPING ESTIMATES OF POTENTIAL DEMAND FOR RENEWABLE WOOD ENERGY PRODUCTS IN ALASKA[*]

*Allen M. Brackley, Valerie Barber
and Cassie Pinkel*

The Forest Service of the U.S. Department of Agriculture is dedicated to the principle of multiple use management of the Nation's forest resources for sustained yields of wood, water, forage, wildlife, and recreation. Through forestry research, cooperation with the States and private forest owners, and management of the National Forests and National Grasslands, it strives—as directed by Congress—to provide increasingly greater service to a growing Nation.

The U.S. Department of Agriculture (USDA) prohibits discrimination in all its programs and activities on the basis of race, color, national origin, age, disability, and where applicable, sex, marital status, familial status, parental status, religion, sexual orientation, genetic information, political beliefs, reprisal, or because all or part of an individual's income is derived from any public assistance program. (Not all prohibited bases apply to all programs.)

---

[*] This is an edited, reformatted and augmented version of the United States Department of Agriculture publication, Forest Service, Pacific Northwest Research Station, General Technical Report PNW-GTR-827, dated August 2010.

ABSTRACT

Brackley, Allen M.; Barber, Valerie A.; Pinkel, Cassie. 2010. Developing estimates of potential demand for renewable wood energy products in Alaska. Gen. Tech. Rep. PNW-GTR-827. Portland, OR: U.S. Department of Agriculture, Forest Service, Pacific Northwest Research Station. 31 p.

Goal three of the current U.S. Department of Agriculture, Forest Service strategy for improving the use of woody biomass is to help develop and expand markets for woody biomass products. This report is concerned with the existing volumes of renewable wood energy products (RWEP) that are currently used in Alaska and the potential demand for RWEP for residential and community heating projects in the state. In this report, data published by the U.S. Department of Commerce, Bureau of Census and the U.S. Department of Energy, Energy Information Agency have been used to build a profile of residential and commercial energy demand for Alaska census tracts. By using peak prices from the fall of 2008, the potential value of a British thermal unit (Btu) from various fuels has been calculated to identify those situations where wood-based fuels are economically competitive or advantageous when compared with alternative fuel sources. Where these situations are identified, the Btu usage has been converted to equivalent volumes of wood energy products. Data have been presented so potential demand is available by census tract. No attempt has been made to define the rate of conversion or the time that it will take for total conversion to renewable wood energy. The ultimate rate of conversion is a function of government policies that encourage conversion, costs associated with converting, and price of alternative fuels. If fuel oil prices increase to the levels experienced in 2008, there would be a strong economic incentive to convert heating systems to use solid wood fuels. If all of the liquid fuels used by the residential and commercial sectors in Alaska were converted to solid wood energy, it is estimated that 1.3 million cords of material would be required annually.

Keywords: Alaska, wood energy, heating fuels.

INTRODUCTION

The current U.S. Department of Agriculture, Forest Service strategy for improving the use of woody biomass (Patton-Mallory 2008) defines four strategy goals. Goals one, two, and four include building partnerships, developing and deploying science and technology, and assuring a supply of biomass. Goal three of the strategy is to "help develop new and expanded markets for bioenergy and biobased products" (Patton-Mallory 2008: 8). All goals are viewed by the Forest Service as important parts of a primary and broader objective of sustaining healthy forests that will survive natural disturbances and threats, including climate change.

A planned short-term action of goal three is to "assist businesses looking to develop new markets or increase the supply of woody biomass products.

A planned short-term action of goal three is to "assist businesses looking to develop new markets or increase the supply of woody biomass products, particularly focused on heating fuels such as pellets and wood chips or commercial use and long-life products that maintain sequestered carbon" (Patton-Mallory 2008: 9). Accordingly, the purpose of this project was to better define the existing profile of heating fuels used in various areas of the state of Alaska and identify opportunities to replace higher cost fossil fuels with renewable wood energy products (RWEP). The need for the estimates of potential markets became noticeable as forest managers, entrepreneurs, civic officials, and citizens expressed increasing interest in using various forms of local biomass directly as sources of energy or for production of energy products (traditional firewood, wood pellets, briquettes, and chips). Although supply was a primary concern, it was assumed that if the material existed, it could be delivered and made available.

Markets, price, and the sources of existing competing energy are all important considerations. Given the high transportation costs to import products to Alaska, local markets and levels of demand are important components of any business plan to produce energy products. Local producers may have a competitive transportation advantage when serving local markets owing to lower transportation costs than those faced by a competitor producing outside the region. It is not uncommon, however, for this competitive advantage in transportation costs to be offset by higher production costs in Alaska. Regardless, the size of the local market and potential demand are of critical importance to firms, organizations, and entrepreneurs interested in production and marketing renewable energy products in Alaska.

There are many factors that will influence the conversion from traditional fossil fuels to RWEPs. Nicholls et al. (2009) reviewed opportunities for the increased use of bioenergy, or RWEP, in the Western United States. They reviewed the legislation and policies that some European nations have adopted to promote conversion and maximum use of RWEP. There are many other factors exterior to government policy that also impact the conversion process in any community or region in the Nation. These include availability of various fuels at the local level; the local cost of the alternative fuels; replacement costs of heating equipment; environmental regulations at the community, state, and federal levels; the existing forest products industry and level of activity; and carbon accounting and environmental economics. A complete analysis of all these problems is beyond the scope of this project. The goal of this project is to provide estimates of the potential demand for RWEP in Alaska, and provide sufficient background material so the reader can begin to estimate the required raw material to meet that demand.

OBJECTIVES

The Alaska Wood Utilization Research and Development Center has a mandate to conduct projects that have the potential to more efficiently utilize the forest resources of Alaska and to promote the economic development of the state. The methodology employed in this project also has the potential to provide researchers with a model that can be applied to any area of the Nation.

The objectives of the project are as follows:

- Provide an overview of the conversion factors and measurement methods— using data generally available from the U.S. Department of Commerce, Bureau of the Census; U.S. Department of Energy, Energy Information Agency (EIA); and other sources, primarily Fairbanks Economic Development Corporation (FEDC)—to compare forms of energy used in the residential and commercial sectors of various geographic regions in Alaska. Special emphasis is placed on measurement and recoverable energy from RWEP, often referred to as biomass products.

- Using Census Bureau, EIA, and FEDC data, estimate the volumes of renewable wood energy that are currently used in Alaska as a primary or secondary fuel for heating purposes.

- Compare the cost per British thermal unit (Btu) of the various alternative sources of energy used in Alaska, and identify situations where renewable wood energy is an economically viable replacement for existing sources of energy.

- Develop estimates of the total volume of renewable wood energy required to replace high-cost alternatives in the residential and commercial sectors in Alaska.

OVERVIEW OF METHODS
TO COMPARE ENERGY USE

There are three major categories of biomass or RWEP: (1) trees harvested specifically for energy (stem or bowl wood limbs, needles, and leaves—including species that will grow under intensive forestry methods), (2) wood fiber residue from sawmills and other plants that process timber (these residues include coarse or chippable residue, sawdust, planer shavings, and bark), and (3) tops and limbs (from trees harvested for traditional products) that are processed in the woods or removed from the woods and converted to energy use. When used for energy, these forms of wood fiber can be burned as is; hogged prior to burning; processed into pellets, compressed logs, or bricks; or as technology improves, manufactured into liquid forms such as biodiesel and ethanol. The focus of this project is the near-term replacement of fossil fuels by the solid forms of RWEP.

> The focus of this project is the near-term replacement of fossil fuels by the solid forms of renewable wood energy products.

Conversion Factors in this Report

The basic energy content of RWEP is reported as Btu values. British thermal unit values are determined by using various calorimeters charged with bone-dry (zero percent moisture content) material and reported in the English system on the basis of weight (Wilson et al. 1987). The standard methods for determining species-specific Btu ratings result in a value referred to as the higher heating value (HHV), or laboratory value, for the material (Briggs 1994, Ince 1979). The reported standard Btu values for wood (Wilson et al. 1987) from the Pacific Northwest extends from a low of about 8,000 Btu/lb (red alder (*Alnus rubra* Bong.))[2] to a high of 9,900 Btu/lb (Alaska yellow-cedar (*Chamaecyparis nootkatensis* (D. Don) Spach)). These values are relatively constant.

Table 1. Higher heating values for wood and bark for species growing in Alaska[a]

Species	Wood	Bark	Percentage of bark by volume
	British thermal units per oven-dry pound		*Percent*
Alaska yellow-cedar	9,900	—	11–13.1
Black cottonwood	8,800	8,882	18.3
Paper birch	8,334	9,900	8–15.7
Quaking aspen	8,200[b]	8,571	8.9–16.5
Red alder	7,995	8,583	13.5
Sitka spruce	8,100	—	—
Western hemlock	8,515	9,421	6.3–16.3
Western redcedar	9,144	8,854	5–13
White spruce	8,890	8,626	8.6–11.7
Average value	8,653	8,977	
Standard deviation	571		

— = no data available.

[a] Some values are shown as the average of the range listed in the source.

[b] This value was obtained by adjusting the value given in air dry condition to oven dry.
   Source: Wilson et al. 1987.

Higher heating values for Alaska species used as fuel are presented in table 1. Using the Wilson et al. (1987) values for nine species that grow in Alaska, an average value for species in the state was calculated as 8,653 Btu/lb. The standard deviation of these values was 571 Btu/lb. The energy value for bark is generally higher than that for wood. Based on this information, the HHV of Alaska wood for the purposes of this analysis is

---

[2] See "Common and Scientific Names" for species names used in this paper.

8,500 Btu/lb. Table 2 contains HHV for other fuels derived from data reported by the EIA (2008d). For wood fuel, the EIA uses a factor of 20 million Btu per cord. For biomass (wood and wood-derived fuels), the EIA uses a factor of 17.2 million Btu per short ton, based on zero moisture content.

Table 2. Higher heating value factors used by the Energy Information Administration (EIA) to calculate reported British thermal unit (Btu) values of selected fuels

Fuel	Btu per barrel	Btu per gallon[a]	Btu per cubic foot	Btu per cord[b]	Btu per short ton[c]	Btu per kilowatt-hour
	Million		*Thousand*	*Million*	*Million*	
Biomass[d]					17.2	
Coal/coke					15.6	
Distillate	5.825	138,690				
Electricity						3,412[e]
Kerosene	5.670	135,000				
Liquefied petroleum gas	3.620	86,190				
Motor gasoline	5.218	124,238				
	Million		*Thousand*	*Million*	*Million*	
Natural gas			1.030			
Wood				20.0		

[a] 42 gallons per barrel.

[b] The cord is a unit of volume that is constant over a wide range of moisture content values. As the moisture content is reduced, the weight of the cord is also reduced while the gross heating value (GHV) per unit of wood weight increases. Total Btu available from the wood is a function of the GHV × weight of wood.

[c] Recoverable Btu are relatively constant over a range of moisture contents for the volume. A Btu value per ton of 17.2 million is based on a bone-dry moisture content (zero moisture content).

[d] Biomass is organic nonfossil material of biological origin constituting a renewable energy source. The EIA defines biomass as "wood and wood-derived fuels."

[e] The Btu per kilowatt-hour conversion is based on the International Units of 3,412.14.

Source: EIA 2008d.

Wood Measurement and Fuel Characteristics

Currently, most wood transported on the highways to forest product facilities is commonly purchased and sold based on fresh cut (green) weight. The cord, however, is still a common unit of measure used to define volumes of wood that are purchased and sold at the retail level for home heating purposes. Numerous sources and forest mensuration texts (Bruce and Schumacher 1950, Evans 2000, Husch et al. 1982) define a cord as a pile of wood, with lengths cut to 4 ft, that has a volume of 128 ft^3 (4 ft high × 4 ft wide × 8 ft long). The unit is further defined as a pile of wood that includes wood, bark, and void air space. The older text (Bruce and Schumacher 1950) was written when the cord unit was a commonly used measure for pulpwood and firewood. This text stated, "There may be anywhere from 60 to 100 cu. ft. of solid wood per cord, depending on the mentioned circumstances" (Bruce and Schumacher 1950: 35). They continued, "Where an average figure is needed, 90 may be considered high and 70 a low, value..." Keep in mind that a cord of wood also includes bark. The exact amount of bark available for burning is a function of species characteristics, time of year the material is processed, handling, and time between harvest and burning. Energy of burnable material may be increased up to 10 percent if all bark is delivered to the stove. Thus, a cord with a wood content of 85 ft^3 would be the equivalent of 93.5 ft^3 when bark is included.

Newer sources (Dunster and Dunster 1996) also make reference to a "face cord" and note that this is material that is cut to usable lengths (12, 16, or 24 in) and, when split for burning and piled, has a face area of 32 ft^2 (or 4 ft high and 8 ft long). A face cord of 24-in material would contain one-half the volume of a standard cord, all other variables being equal.

The previous paragraphs start to identify some of the problems incurred when energy values derived on a weight basis are applied to a product that is measured by volume. The previous comments are concerned with the problems of measuring the unit volume, but the magnitude of the problem becomes clearer when the weight-based energy values of the material are applied to volumes.

A review of the USDA Forest Service wood handbook (USDA FS 1999) indicates that the reported green specific gravity of the wood species (softwoods and hardwoods) that grow in the United States range from a low of 0.29 to a high of 0.66 (USDA FS 1999: table 4-3b). Conversion of these values to density results in values that range from a low of 18.1 lb to a high of 41.2 lb of green wood per cubic foot of wood material. A similar review of specific gravity values for material at 12 percent moisture content shows that the specific gravity of this relatively dry material ranges from 0.31 to 0.75, or a range of density from 19.3 to 46.8 lb/ft^3. Note that green material will shrink when dried below fiber saturation point and will have less volume, but the impact of this volume change is minimal when compared with the natural variation in cord weights composed of various species. In the above comments relative to Btu values, it was noted that for the purposes of this report,

all RWEP would be assigned an HHV rating of 8,500 Btu/lb. Once this starting point was established, the next step required to calculate the recoverable Btu of a heating system was to adjust for fuel moisture content (Briggs 1994, Ince 1979). This calculation takes into account the energy that is required to drive off the water during the combustion process.

As a rule of thumb, many consultants assume that fresh-cut wood is half (50 percent) water.

The moisture content of fresh-cut wood differs considerably among species (Bowyer et al. 2003). In general, the heartwood is of lower moisture content than the sapwood. As a rule of thumb, many consultants assume that fresh-cut wood is half (50 percent) water. Wood is hydroscopic in nature, and it gains and looses moisture with the surrounding environment. In most coastal areas of Alaska, green wood, stored so that it is covered from the rain and piled so that air will circulate through the material, will dry to an equilibrium moisture content of 15 to 16 percent, green basis. Inland areas will reach equilibrium at slightly lower levels. Drying rates of Sitka spruce (*Picea sitchensis* (Bong.) Carr.) and western hemlock (*Tsuga heterophylla* (Raf.) Sarg.) logs were reported by Nicholls and Brackley (2009). Regardless, in situations where older heating equipment is used, wood at the stated equilibrium moisture contents may create unacceptable levels of emissions and exceed Environmental Protection Agency (EPA) emission standards (ECFR 2009). The problem may be especially acute during periods of temperature inversion, which happens frequently in the winter in Fairbanks and Juneau, the second and third largest cities in Alaska. In more urban areas of Alaska, local ordinance may require heating equipment that meets EPA emission standards or limits the use of older stoves during certain weather conditions.

Most moisture content values in the Forest Service wood handbook (USDA FS 1999) are based on the dry weight of the wood (dry basis). In energy applications, adjustments for moisture content are based on the original green weight of the material (green basis). The following formula allows conversion from dry to green basis when the values are expressed as a percentage (Kollman and Cote 1968):

$$MC_{db} = (100 \times MC_{gb}) / (100 - MC_{gb}) \text{ or}$$
$$MC_{gb} = (100 \times MC_{db}) / (100 + MC_{db}),$$

Where

MC_{db} = moisture content, dry basis expressed as a percentage, and
MC_{gb} = moisture content, green basis expressed as a percentage.

Once a moisture content of 12 percent dry basis is adjusted to 10.7 percent green basis, the equation presented by Briggs (1994) or Ince (1979) can be used to calculate gross heating value (GHV) when burning takes place, using the following formula:

$$GHV = HHV \times (1 - [MC_{gb} / 100]),$$

Where

GHV = gross heating value,
HHV = higher heating value, and
MC_{gb} = moisture content, green basis expressed as a percentage.

Thus, the HHV value of 8,500 Btu/lb for wood material adjusted using the above formula (with MC_{gb} equal to 10.7) will produce a GHV of 7,590 Btu/lb. When the range of cubic-foot weights (19.3 to 46.8 lb/ft^3) are applied to the calculated GHV value, the GHV of a cubic foot of wood ranges from a low of 146,487 Btu/ft^3 to a high of 355,212 Btu/ft^3. Cords represent volume that is expressed in terms of cubic feet.[3] In this analysis, we have considered volumes that have been reduced to a moisture content of 10.7 percent green basis. If the analysis were repeated and values of GHV at higher moisture contents included, the range would more than double. This analysis is intended to demonstrate some of the problems of assigning weight-based energy HHV values to any volume-related factor. The bottom line is to always start the process with weight values and then make the conversion to volume in the

---

[3] The EIA (2008d) value for wood in table 2 (20 million Btu/cord) can be converted to 235,294 Btu/ft3 (assuming 85 ft3 per cord).

final stages. When the final conversion from weight to volume is made, always provide the factor for the conversion.

Up until this point, the focus of energy recovery has been on the characteristics of the fuel. Ultimately, the final calculations to determine the recoverable heat and combustion efficiency of any heating system must also take into consideration the equipment that is used and the insulation of the building that is being heated.

> In any heating system, the energy from the combustion process is either vented up the chimney (stack heat loss), lost to the environment during transfer to the area being heated, or applied to the area being heated.

Efficiency of Heating Systems

In any heating system, the energy from the combustion process is either vented up the chimney (stack heat loss), lost to the environment during transfer to the area being heated, or applied to the area being heated. The direct transfer of heat (energy) may be by a process, singularly or in combination, of radiation and convection. Both Briggs (1994) and Ince (1979) provided formulas for converting GHV values to recoverable heat or combustion efficiency values.

The product and marketing literature for space heating stoves (regardless of fuel type), furnaces, water heaters, and other appliances include an efficiency rating expressed as a combustion efficiency value. Most modern wood heating units also include an EPA-based emissions rating. The standards that products must meet are defined by the National Appliance Energy Conservation Act (NAECA 1987) and Department of Energy regulations. Technically, the efficiency rating reported by Briggs (1994) is comparable to the ratings reported for various heating stoves under the NAECA. Some sources present tables where HHV values for various systems are reduced by the efficiency rating of the equipment to obtain estimates of combustion efficiency or deliverable Btus for various fuels. The results of this approach are reasonable for simple systems using most liquid fuels with minor or low moisture contents. They do not, however, provide reasonable estimates for fuels such as wood, many RWEP, and Alaska coal with high moisture contents. Gross heating values reduced by appliance efficiency are probably optimistic and higher than most installations, given that stove testing is done under optimal conditions where stack properties and air feeds are controlled. The installation of the product in less controlled conditions of a specific home will probably result in a combustion efficiency value lower than that reported by the manufacturer.

Impact of Climate on Home Heating

From north to south, the state of Alaska is a distance of about 1,000 mi. Within the state, the mean annual heating degree days using a base of 65 °F, range from a low of 6,855 at Annette Island in southeast Alaska to a high of 19,719 degree days in Kuparuk, a community on Prudhoe Bay in northern Alaska (ACRC 2009). Degree days provide a rough estimate of heating requirements. The values may not, however, be directly correlated to energy use for heating, unless the characteristics of housing units (size of heated living area, amount of insulation, efficiency of heating systems, etc.) in each location are taken into consideration. A poorly insulated house in a warmer area with fewer degree days may require more energy for heating than a well-insulated house in a colder area with more degree days.

> A poorly insulated house in a warmer area with fewer degree days may require more energy for heating than a well-insulated house in a colder area with more degree days.

Review of U.S. Census Bureau Data

The federal government conducts a census every 10 years to determine the apportionment of congressional representatives among states. Although the government collects the data for apportionment, distribution of federal funds, delineation of legislative districts, and other purposes, the information is also available to the general public and businesses. The information is especially valuable to businesses for use in marketing applications.

As stated in the objectives, one purpose of this project is to review the current profile of energy used to heat residential homes and commercial businesses in Alaska. An important statistic for use in this activity is housing units in census tracts.

Currently, the methods for collecting census housing information are in a state of flux as the Census Bureau converts from periodic to continuous systems of data collection (U.S. Census Bureau 2000, 2004, 2005a, 2006a). During the transition period, updated estimates of the total number of housing units in Alaska (or any state) are available for 2006, but many of the tables for census tracts are not updated to reflect these changes. For the purposes of this project, updated census tract housing totals for 2006 have been used to update the original 2000 table, based on the assumption that change in all units is proportional. This adjustment required minimal effort and allowed the authors to proceed with preliminary calculations relative to local demand and may have induced minor but insignificant errors in the final results.

A map showing the location of Alaska census tracts is presented in figure 1.

The American Community Survey (U.S. Census Bureau 2006a) also collects information relative to home living area, unit age, occupancy, and home ownership. In the following sections, the above data may be modified to reflect estimates of occupied homes, owner-occupied homes, or other factors. Such modifications have been noted as they appear. The profile of energy sources used for home heating on a household basis in Alaska from the Census Bureau sources are presented in figure 2. It shows that natural gas (45 percent) is the primary fuel used for home heating in Alaska, and wood is used for heat in 4 percent of the homes.

Natural gas (45 percent) is the primary fuel used for home heating in Alaska, and wood is used for heat in 4 percent of the homes.

Figure 1. Alaska census tracts (U.S. Census Bureau 2005b).

Review of Energy Information Agency Data

The EIA State Energy Data System (SEDS) reports energy consumption estimates for each state. The basic data on energy sources (EIA 2008a, 2008b) are reported in terms of Btu values and units of sale used in the commerce of the United States. The conversion factors used to calculate Btu are reported in the technical notes and documentation to SEDS (EIA 2008d). The information (EIA 2008a) is reported for the following sectors of the economy: residential, commercial, industrial, transportation, and electric power. The profile of Btu usage for Alaska's residential sector based on EIA energy sources are reported in figure 3. The overall breakdown is similar to that shown in figure 2, with natural gas having the highest percentage (46) and biomass (or wood and wood-derived fuels) at 7 percent.

The EIA data (fig. 3) report the energy profile for all residential uses (heating, appliances operation, lighting, and others). The Census Bureau data (fig. 2) report the profile for heating based on number of households. The definition used by EIA to define the residential sector does not correspond directly with the Census Bureau-defined sector. The major difference between the two classification systems is that EIA considers multiunit apartment buildings as part of the commercial sector, whereas the Census Bureau considers these units as part of the residential component. Given the rural nature of Alaska, however, it is considered a minor problem.

Table 3 presents the profile of energy use for Alaska. With respect to forms of fuel used, the Alaska energy profile can be characterized as one where over twice the volume of natural gas (54.3 percent vs. 22.6 percent) and less than one-tenth the volume of coal (1.8 percent vs. 22.7 percent) are used compared to national averages (EIA 2008a). The energy profile also shows that an unusually high percentage of jet fuel is used in Alaska (22.7 percent of state energy use as opposed to a 3.5 percent average nationally) (EIA 2008a). Table 3 also shows the power generated from fossil fuels that is assigned to the residential, commercial, and industrial sectors as reported by the EIA. In the EIA tables (EIA 2008a), total energy use is reported, and to prevent double accounting, fuel oil, natural gas, and coal used to generate electricity are assigned to the electrical sector. This material is converted to electricity and sold to the other sectors. The adjusted table accounts for all the energy consumed by those three sectors, including power generated by thermo mechanical systems (electricity generated by using fossil fuel).

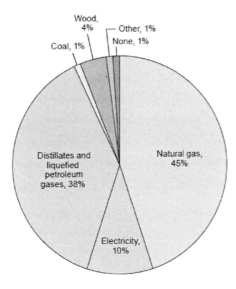

Figure 2. Fuel used as primary source for home heating in Alaska (2005) based on number of reported housing units by Bureau of Census 2000 (U.S. Census Bureau 2000, 2005a, 2006b).

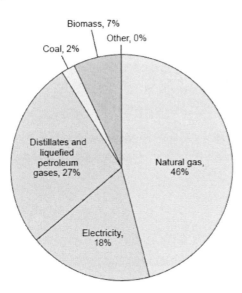

Figure 3. Energy use in the residential sector of Alaska (2005) based on British thermal unit values. Biomass includes wood and wood-derived fuels (EIA 2008a).

One final comment about the EIA data: EIA energy values are derived from sales data and other sources. The survey methods for collecting estimates of wood consumed within the state are in part derived by combining Census Bureau and EIA survey information (EIA 2008d).

> The Alaska energy profile can be characterized as one where over twice the volume of natural gas (54.3 percent vs. 22.6 percent) and less than one-tenth the volume of coal (1.8 percent vs. 22.7 percent) are used compared to national averages.

Additional Data

The Alaska Housing Manual (AHFC 2000) provides specific information relative to existing heating systems used in Alaska. This source reports that forced-air furnaces are very popular in the Anchorage-Valley area of Alaska, but fairly rare in other areas of the state. In other regions of Alaska, hydronic heating systems with baseboards or radiant floor systems are the standard. This source makes reference to space heaters produced by Toyo, Monitor, and Rinnai,[4] but provides no information relative to level of use.

In 2007, the FEDC completed a survey to assess community and consumer interest in use of pellet fuel (Robb 2007). Respondents were from the three major population areas of the state, including greater Fairbanks, central Alaska (Anchorage, Valley, and Kenai), and southeast Alaska. Survey questions were designed to collect information relative to current fuel types being used, current levels of energy use in both primary and secondary systems, fuel price, and the characteristics of existing wood-burning equipment. The updated Census Bureau household information can be used with FEDC survey data to obtain updated estimates of volumes of renewable wood energy being used for secondary heating in the various regions of Alaska.

The FEDC survey represented information supplied by homeowners, a subset of total housing units. The reported usage of the major source of energy (fuel oil in Fairbanks and natural gas in Anchorage) for home heating was higher than those reported for all housing units in the Census Bureau housing survey. This trend in the data suggests that the source of heating energy for

---

[4] The use of trade or firm names in this publication is for reader information and does not imply endorsement by the U.S. Department of Agriculture of any product or service.

rental units may differ from that of owned units. The authors were unable to locate any data relative to rental units that would confirm this speculation.

Table 3. Sources of energy consumed by various segments of the Alaska economy, 2005

Source	Alaska distribution	Residential sector	Commercial sector	Industrial sector	Transport sector[a]	Electric power	Total
	Percent	*Trillion Btu*					
Asphalt/ roads	0.2	0	0	1.2	0	0	1.2
Aviation	.2	0	0	0	1.4	0	1.4
Biomass	.4	2.8	0.5	0.1	0	0	3.4
Coal	1.8	0.7	7.2	0	0	6.1	14.0
Distillate fuel	9.2	9.4	5.9	11.1	43.7	3.1	73.2
Ethanol	0	0	0	0	0.5	0	0.5
Hydroelectric	1.8	5.1	6.7	2.8	0	0	14.6
Jet fuel	22.7	0	0	0	181.1	0	181.1
Kerosene	0	.2	0	0	0	0	.2
Liquefied petroleum gas	.1	.8	.1	0	0	0	.9
Lubricants	.1	0	0	.1	.5	0	.6
Motor gas	4.5	0	.9	.5	34.3	0	35.7
Natural gas	54.3	18.1	16.9	356.7	2.7	39.5	433.9
Other	.0	.1	.1	0	0	0	.2
Other petroleum	4.3	0	0	34.3	0	0	34.3
Residual fuel	.6	0	0	0	.1	4.4	4.5
Total[b]	100	37.2	38.3	406.8	264.3	53.1	799.7

Table 3. (Continued)

Source	Alaska distribution	Residential sector	Commercial sector	Industrial sector	Transport sector[a]	Electric power	Total
	Percent	*Trillion Btu*					
Fossil electricity consumed		1.9	2.5	1.10			
Total energy consumption		39.1	40.8	407.9			
Electricity sold to sector[c]		7.0 (34.8%)	9.2 (45.8%)	3.9 (19.4%)			20.1 (100%)

[a] Includes passenger vehicles.

[b] This total does not take into account transfers of electricity from the power sector to the other sectors. Thus, electricity from the power sector was added to the residential, commercial, and industrial sectors to account for all the energy consumed by those sectors.

[c] Electricity sold in Alaska is generated both by hydro sources and fossil fuel plants. The generation of electricity from fossil fuels is very inefficient. Fossil fuels consumed by the electric power sector are converted to electricity and sold to the other sectors. Minor amounts of electricity are utilized by the electric power sector and the amounts of energy leaving the generating plants are subject to transmission loss, prior to being sold to the various other sectors of the Alaska economy.

Btu = British thermal unit.

Note: The table contains rounding errors owing to the energy amounts in the source being in units of 1 trillion Btu.

Source: EIA 2008a: tables 7, 8, 9, 10, 11, and 12.

ESTIMATES OF CURRENT USE OF RWEP FOR HOME HEATING IN ALASKA

Combining the data from the Census Bureau, EIA, and FEDC allows calculation of the volumes of wood currently being used for primary and secondary heating in Alaska census tracts. This was accomplished by using a three-stage process. First, Census Bureau 2000 data were updated by using 2006 Census QuickFacts (U.S. Census Bureau 2006b) data to reflect the number of housing units using distillate fuels (fuel oil, kerosene, and liquefied petroleum gas) for heating. The housing numbers were adjusted by removing all apartment units that were in buildings with five or more units. The housing units were also adjusted to reflect occupancy (half of unoccupied units were considered as unheated) and the smaller size of mobile homes (half of mobile homes were subtracted). This provided an adjusted estimate of the number of household units in each census track that were using distillate fuels.

Second, by using the EIA energy values (EIA 2008a) for residential use of distillate fuels and the adjusted number of housing units from step 1, average Btu usage for a unit was calculated. Given the temperature extremes that exist in the state, the authors were reluctant to apply this average to the census tracts without some weighting factor to account for the increased number of heating days and severity of winters in northern areas of the state.

The energy required to heat a house for a winter season is based on a combination of factors (e.g., quality of construction and insulation, size of the living area being heated, severity of the climate, fuel used, efficiency of the heating system), but a complete analysis of all of these factors was beyond the scope of this project. To simplify the analysis, it was assumed that all of these factors, other than climate, would be averaged. The impact of climate would be reflected by degree days using 65 °F as the basis (ACRC 2009).

Third, by using information from the Alaska Climate Research Center (ACRC 2009), the census tracts were assigned a reported degree-day value for a community within each census tract. Given that airports have a constant need for current weather information, an attempt was made to use airport locations whenever possible. If an airport location was not reported, a named community was used.

Given an estimate of the number of housing units and a weighting value to account for climate, the average house Btu value from step 2 was prorated to each census tract by using the number of housing units times number of degree days.

The resulting estimates of annual Btu requirement for heating are presented in table 4. It shows the average for the state as almost 113 million Btu per home.

Estimated volumes of firewood consumed annually in Alaska were based on household Btu values derived from EIA, Census Bureau, and expanded FEDC data. The results for the major population centers in the state are reported in table 5. In the calculations, the Btu value assigned to the firewood assumed that the material had a moisture content of 50 percent and it was burned in a unit with a combustion efficiency of 60 percent. Using procedures outlined in Briggs (1994), such material would produce approximately 2,550 Btu/lb. The weight of the same material dried to 20 percent moisture content would be approximately 62 percent of green weight and have approximately a recoverable Btu value of 5,000 Btu/lb.

Table 4. Annual British thermal units (Btu) required for home heating in Alaska areas defined by Census Bureau tracts (derived from usage of distillate fuels)

Region	Borough/census area	Equivalent number of housing units	Btu per home
AA	Aleutians S. Borough	544	96,125,892
	Aleutians W. Borough	1,386	89,504,797
	Nome	2,786	136,762,610
	North Slope Borough	895	195,772,366
	Northwest Arctic Borough	1,825	155,655,733
	Wade Hampton Census Area	1,686	127,711,114
	Yukon-Koyukuk Census Area	2,135	146,074,149
	Total and average	11,257	137,140,799
GA	Anchorage	3,940	104,717,312
	Kenai Peninsula	10,213	111,828,488
	Matanuska-Susitna	9,080	106,087,539
	Valdez-Cordova	3,226	94,485,621
	Total and average	26,460	106,684,892
GF	Denali	806	131,081,671
	Fairbanks North Star	19,922	139,823,116
	Southeast Fairbanks	1,757	148,144,492
	Total and average	22,485	140,159,841
SE	Haines	873	86,394,283
	Juneau	7,083	85,754,177
	Ketchikan	4,072	71,561,831
	Prince of Wales-Outer	1,800	71,561,831
	Sitka	2,432	72,241,943
	Skagway-Hoonah-Angoon	1,515	86,444,291
	Wrangell-Petersburg	2,298	77,072,742
	Yakutat	347	92,395,275
	Total and average	20,421	79,278,059

Region	Borough/census area	Equivalent number of housing units	Btu per home
SW	Bethel	4,180	127,711,114
	Bristol Bay	633	111,058,360
	Dillingham	1,749	114,078,860
	Kodiak Island	3,892	88,634,653
	Lake and Peninsula	1,029	111,058,360
	Total and average	11,483	109,980,779
State	Total and average	92,106	112,913,618

Note: AA = Aleutians and Arctic, GA = greater Anchorage, GF = greater Fairbanks, SE = southeast, and SW = southwest.

Given the above procedures and conversion factors, it is estimated that there are 8,632 homes using wood as the primary source of heat in Alaska (table 5). The total volume of firewood used annually as the primary source of heating is 76,203 cords. Given the FEDC survey, it was also possible to estimate the number of homes (31,227 units) and volume of firewood (53,502 cords) used in secondary heating.

In total, it is estimated that about 138,656 cords of wood are used as a primary and secondary source of heat in Alaska. It might be assumed that the firewood removals from the Alaska forests are equal to the same number. This, however, is not necessarily the case. The relationship between volume used and harvest is discussed later in this report.

An estimated 8,632 homes in Alaska use wood as their primary source of heat. The total volume of firewood used annually for heating is 76,203 cords.

Table 5. Estimated volumes of firewood consumed annually in Alaska[a]

Region	Number of homes using wood as primary source of heat	Primary heat, firewood	Primary heat, green wood	Primary heat, green wood	Number of homes using wood as secondary source of heat	Secondary heat source volumes[b]	Total volume of primary and secondary usage
		Pounds	Tons	Cords		Cords	Cords
Aleutians and Arctic	1,337	75,541,179	37,771	15,108	2,612	5,225	20,333
Greater Anchorage	3,578	144,385,804	72,193	28,877	18,103	36,205	65,082
Greater Fairbanks	1,590	88,923,133	44,462	17,785	4,755	9,510	27,295
Southeast	1,698	52,098,096	26,049	10,420	3,894	7,787	18,207
Southwest	430	20,066,712	10,033	4,013	1,863	3,726	7,739
Total	8,632	381,014,924	190,507	76,203	31,227	62,453	138,656

[a] Based on household British thermal unit values derived from Energy Information Administration distillate fuels, Census Bureau housing data, and expanded Fairbanks Economic Development Corporation (FEDC) survey volumes. It was assumed that the firewood had a moisture content of 50 percent and it was burned in a unit with a combustion efficiency of 60 percent, producing about 2,550 Btu/lb.

[b] Based on FEDC pellet fuel survey, with a median value of two cords per owner-occupied housing unit. The Aleutians and Arctic, and Southwest census tracts were not included in the FEDC survey. The secondary use percentage from Greater Fairbanks was applied to the Aleutian and Arctic region and the Southeast rate to the Southwest region to obtain the estimates above. No estimate for secondary firewood use in rental units is included.

Note: All green weights have been converted to cords based on a factor of 2.5 tons per cord. Use of this factor will underestimate softwood volumes and overestimate hardwood volumes.

Sources: EIA 2008a; Robb 2007; U.S. Census Bureau 2000, 2006b.

COST COMPARISON FOR
ALTERNATIVE ENERGY SOURCES

From the consumer's point of view, any economic assessment of fuels and sources of energy must provide an answer to the basic question, "What will each alternative cost?" There are two components of cost. First is the cost of purchasing equipment (fixed costs) for a specific fuel option and second is the cost for annual maintenance and the fuel itself (variable). Although the question seems simple, the process of developing the answer, given the maze of measurement units used in the commerce of fossil and renewable forms of energy, is complex.

Given the hydroscopic nature of wood, an understanding of moisture content and its impact on recoverable energy is critical to this report. It must be recognized that all fuels (e.g., coal, gas, oil, and wood) may include moisture that must be removed in the combustion process. In gas and oil, the moisture content is most commonly minimal and the efficiency of the burning equipment itself results in a loss of energy available for heating or powering a process. The most logical way to compare cost of alternative sources of energy is to compare the cost of the Btu recoverable from each product. In this project, the recoverable Btu value of selected alternative sources of energy were calculated to show the break-even price that could be paid for the alternative.

The commerce of wood and the units of measurements therein are poorly understood by the general public. Many individuals in the forest products industry are just becoming exposed to the conversion factors and evaluation of biomass-related energy products. In many areas, costs for purchasing standing trees, harvesting, and delivery to mills is readily available. In such locations, mills procuring wood advertise prices they are willing to pay for specific species and products. In many heavily forested states, information relative to standing timber and delivered values of forest products are collected and published by university extension agencies and state-supported marketing programs. Such is not the case in Alaska.

Another complication associated with Alaska is that almost all of the timber harvest in the region processed by sawmills is of high quality and large size. The mills produce residual products (chips, sawdust, bark, or mixtures thereof) that can be used as a source of energy, but Alaska lacks mills that process timber directly from the round-log form to fiber. Given these Alaska-related problems, the following is a review of timber economics in general. It has been assumed that harvesting for energy products would require low-quality logs. Generally available information from Oregon (ODF 2009), an area whose conditions and timber species are somewhat comparable to those of southeast Alaska, has been used to develop a hypothetical minimum price for low-grade logs and young growth harvested in Alaska.

An explanation of selling prices of wood can best be reviewed by commonly relating them to the following two situations. In the first case, a landowner that owns standing timber may sell the material to a logger or mill as standing timber. In this instance, the value of the wood is referred to as "stumpage value." The logger or the mill owner has the right to enter onto the land, harvest the timber, and deliver it to the mill. The logger pays the landowner for the value of timber harvested, or "stumpage." The costs of harvesting and transportation to the mill are not included in the "stumpage value." In Alaska, stumpage may be purchased from the U.S. Forest Service, several state agencies, Native landowners, and occasionally in small volumes from private landowners.

Moving along the chain of commerce to the mill owner that produces products from logs, the cost of the material delivered to the mill includes the stumpage value, all in-woods harvesting costs, cost of trucking or transportation to the mill, and any overhead costs for management, supervision, and scaling of material. The delivered cost is synonymous with the term "pond value."[5] An excellent explanation of various log values is available at the Oregon Department of Forestry (ODF) Web site http://www.oregon.gov/ODF.

It is certain that costs and prices in Alaska will be higher than anything reported from the ODF (2009) site. But as an example, the average pond price for low-grade logs in the fourth quarter of 2008 from this source (ODF 2009) is reviewed. The average price for 46 sales of low-grade western hemlock and spruce (*Picea* spp.) logs was calculated as $271 per thousand board feet (mbf). A review of selected sales from state lands for the same period indicated a woods-run stumpage price for all grades of spruce logs of approximately $100/mbf and slightly less than $50/mbf for hemlock. As stated, this is an average price for all harvested grades. The prorated stumpage value for low-grade logs could be 25 to 50 percent of the average values. If these prices are applied to the pond price, dollar amounts paid by the mill owner would be distributed in the following manner, assuming the end product was low-grade

---

[5] The term "pond value" is a holdover from the days when logs were stored in water once delivered to the mill. Water-stored logs were protected against log-boring insects and, in the days before mechanized equipment, easy to move from storage by floating them to the mill.

hemlock. Approximately \$25/mbf would be paid to the landowner to cover the cost of the wood. An amount of \$232.50/mbf would be paid to the contractor delivering the logs to cover harvesting, trucking, in-woods supervision, and hopefully some amount for profit and risk. A small amount, estimated at approximately 5 percent for the purposes of this example or \$13.50/mbf, would go to cover the cost of scaling and yard handling.

Firewood is traditionally sold by the cord and the final question that must be addressed is, "Given a pond price of \$271/mbf, what would be the value per cord?" First, the material purchased at the stated rates is low grade and tends to be in logs with small scaling diameters (less than 12-in diameter) and the ratio of cords to thousand board feet in such material is high. The ratio is also extremely sensitive to the length of the scaled logs. Based on experience and data from two young-growth projects conducted by the Alaska Wood Utilization Center[6] (Brackley et al. 2009b, Nicholls and Brackley 2009), the conversion factor could range from 2.5 to 4.5 cords/mbf of scaled material. The cord price for this range would be between \$108.40 and \$60.22 per cord of unprocessed wood (not cut to length or split).

Table 6 presents information relative to the dollar value of a therm of energy (a therm is defined as 100,000 Btu of energy) from various alternative fuels. In table 7, the value of selected RWEP (green cords, dry cords, tons of green material, tons of dry material, and tons of pellets) is shown based on the dollars per therm from alternative sources. The tables have been prepared so the reader can reference the costs of the alternatives in terms of the commonly utilized units of measurement. To interpret the tables, using fuel oil as an example, if the price per gallon is \$4.00, then the value per therm (100,000 Btu) is \$2.88 (see table 7). If the fuel were burned in a unit that was 85 percent efficient, the cost per therm would be adjusted to \$3.39. The value of a cord of wood at 50 percent moisture content, burned in a stove with 50 percent efficiency, would provide 106.3 therms of energy. Based on the cost of \$3.39 per Btu for oil, a person could pay up to \$361 for a cord of wood to replace oil at \$4.00/gal (see table 7). Interpretation of the data at this point will depend on the self-motivation of the individual user and their willingness to invest in equipment and cut, transport, and split wood. A willing user that has a source of wood on the stump can reduce energy cost to a fraction of the above amounts.

---

[6] Nicholls, D.; Brackley, A.M. 2008. House log drying rates in southeast Alaska for covered and uncovered softwood logs. Unpublished data. On file with: USDA Forest Service, Pacific Northwest Research Station, Alaska Wood Utilization Research and Development Center, 204 Siginaka Way, Sitka, AK 99835.

Table 6. Value of selected energy alternatives based on assigned value (dollars) per therm of energy

Energy alternative (unit)	Assigned value of 1 therm of energy[a]										
	$2.00	$2.25	$2.50	$2.75	$3.00	$3.25	$3.50	$3.75	$4.00	$5.00	$6.00
	Dollars per unit										
Fuel oil (gallon)	2.77	3.12	3.47	3.81	4.16	4.51	4.85	5.20	5.55	6.93	8.32
Propane (gallon)	1.72	1.94	2.15	2.37	2.59	2.80	3.02	3.23	3.45	4.31	5.17
Natural gas (Mcf)	20.60	23.18	25.75	28.33	30.90	33.48	36.05	38.63	41.20	51.50	61.80
Electricity (kW)	0.07	0.08	0.09	0.09	0.10	0.11	0.12	0.13	0.14	0.17	0.20
Wood pellets (ton)	340	383	425	468	510	553	595	638	680	850	1,020
Wood 50%MC (cord)	425	478	531	584	638	691	744	797	850	1,063	1,275
Wood 20%MC (cord)	425	478	531	584	638	691	744	797	850	1,063	1,275
Wood 50%MC (ton)	170	191	213	234	255	276	298	319	340	425	510
Wood 20%MC (ton)	272	306	340	374	408	442	476	510	544	680	816

[a] A therm of energy equals 100,000 British thermal units (Btu). Given the assigned value per therm, each row presents the dollar value of an alternative fuel in terms of the units commonly used in commerce as the basis for selling and purchasing the energy product. The values in each column represent the maximum price that a consumer can pay for the alternative. If the market price for an alternative is lower than the listed value, there is an economic incentive for the consumer to change to that energy source. When considering changes in energy source, the consumer must also take into consideration the cost (investment) for equipment upgrades and the conversion efficiency of the alternative.

Note: All references to moisture content (MC) are green basis. The recoverable Btu from a volume of wood is constant over a range of moisture contents. As a volume of wood dries, the weight of the material decreases but the gross heating value (GHV) of the material increases. Weight × GHV = a constant value.

Table 7. Dollar value of British thermal units (Btu) in selected renewable wood energy products based on Btu value of alternative sources of energy

Wood fuel and burning equipment characteristic				Alternative source of energy to be replaced			You can pay up to:
Unit of commerce	Moisture content	Burning unit efficiency	Recoverable energy	Cost	Heating Unit efficiency	Adjusted cost for recoverable energy	Cost per unit to replace alternative
	Percent		Therm[a]	Dollars per therm	Percent	Dollars	
Fuel oil at $2.00 per							
Green cord	50	50	106.3	1.44	85	1.70	180
Dry cord	20	50	106.3	1.44	85	1.70	180
Green ton	50	80	68.0	1.44	85	1.70	115
Dry ton	20	50	68.0	1.44	85	1.70	115
Dry ton	20	80	108.8	1.44	85	1.70	185
Pellets ton	6.5	80	127.2	1.44	85	1.70	216
Fuel oil at $4.00 per							
Green cord	50	50	106.3	2.88	85	3.39	361
Dry cord	20	50	106.3	2.88	85	3.39	361
Green ton	50	80	68.0	2.88	85	3.39	231
Dry ton	20	50	68.0	2.88	85	3.39	231
Dry ton	20	80	108.8	2.88	85	3.39	369
Pellets ton	6.5	80	127.2	2.88	85	3.39	432
Natural gas at $10 per thousand cubic feet:							
Green cord	50	50	106.3	0.97	90	1.08	115
Dry cord	20	50	106.3	0.97	90	1.08	115
Green ton	50	80	68.0	0.97	90	1.08	73

Table 7. (Continued)

	Wood fuel and burning equipment characteristic				Alternative source of energy to be replaced			You can pay up to:
Unit of commerce	Moisture content	Burning unit efficiency	Recoverable energy	Cost	Heating Unit efficiency	Adjusted cost for recoverable energy	Cost per unit to replace alternative	
	Percent	Percent	Therm[a]	Dollars per therm	Percent	Dollars		
Dry ton	20	50	68.0	0.97	90	1.08	73	
Dry ton	20	80	108.8	0.97	90	1.08	118	
Pellets ton	6.5	80	127.2	0.97	90	1.08	137	
Natural gas at $15 per thousand cubic feet:								
Green cord	50	50	106.3	1.46	90	1.62	172	
Dry cord	20	50	106.3	1.46	90	1.62	172	
Green ton	50	80	68.0	1.46	90	1.62	110	
Dry ton	20	50	68.0	1.46	90	1.62	110	
Dry ton	20	80	108.8	1.46	90	1.62	176	
Pellets ton	6.5	80	127.2	1.46	90	1.62	206	
Electricity at $0.10 per kilowatthour:								
Green cord	50	50	106.3	2.93	90	3.26	346	
Dry cord	20	50	106.3	2.93	90	3.26	346	
Green ton	50	80	170.0	2.93	90	3.26	554	
Dry ton	20	50	68.0	2.93	90	3.26	221	
Dry ton	20	80	108.8	2.93	90	3.26	354	
Pellets ton	6.5	80	127.2	2.93	90	3.26	414	
Electricity at $0.15 per kilowatthour:								
Green cord	50	50	106.3	4.40	90	4.88	519	

Wood fuel and burning equipment characteristic				Alternative source of energy to be replaced			You can pay up to:
Unit of commerce	Moisture content	Burning unit efficiency	Recoverable energy	Cost	Heating Unit efficiency	Adjusted cost for recoverable energy	Cost per unit to replace alternative
	Percent		Therm[a]	Dollars per therm	Percent	Dollars	
Dry cord	20	50	106.3	4.40	90	4.88	519
Green ton	50	80	170.0	4.40	90	4.88	830
Dry ton	20	50	68.0	4.40	90	4.88	332
Dry ton	20	80	108.8	4.40	90	4.88	531
Pellets ton	6.5	80	127.2	4.40	90	4.88	621

[a] A therm is a term used to describe 100,000 British thermal units of energy.

Given the above analysis, a landowner or logger in the business of selling firewood can charge up to the stated amount of the alternative for cut-to-length and split material and be competitive with the alternative. If they can reduce their price to less than the above amount and are satisfied with the profit margin, they have a competitive advantage over the alternative at the stated price. The lower the price, the more competitive they become and the more incentive the user has to make an investment in new heating equipment (capital costs required to replace fuel systems with RWEP could be very high) to convert from the alternative to RWEP. If the cost of delivering the material to a wood processing yard is $108.40, the high end of the estimated delivered cost, there is still a tremendous opportunity ($252.60 obtained by subtracting the $108.40 per cord price for delivery to the wood yard from $361 alternative cost of oil) to make a profit from the activity.

POTENTIAL DEMAND FOR RWEP FOR HOME HEATING IN ALASKA

Given this analysis, it is concluded that at $3.00/gal for fuel oil there is a price incentive for users of distillate fuels to convert to RWEP. With recent electricity prices over $0.10/kWh, there is also a price incentive for consumers using this form of energy to convert. For the purposes of this report, it is assumed that the maximum potential is in fact best defined as the Btu level used by consumers using distillate fuels. Estimates of the maximum potential volumes of RWEP that might be required annually to meet this level of demand are shown in table 8.

   At $3.00/gal for fuel oil, there is a price incentive for users of distillate fuels to convert to renewable wood energy products.

In table 8, the 2006 updated Census Bureau data have been adjusted so that the number of housing units more closely corresponds to the EIA-defined residential sector. The adjustments are the same as reviewed previously in the report. Data for EIA residential distillate fuel use was then prorated to the number of housing units, weighted by the number of degree days for a named location in the census tract. Once the census tract distillate Btu value was available, it was converted to an equivalent Btu of RWEP assuming that the wood was dried to 20 percent moisture content and burned at a combustion efficiency of 60 percent.

Table 8. Annual volume of wood or pellets required to replace distillate fuel used in residential and commercial sectors in Alaska census tracts

Borough/census area	Adjusted number of housing units	Residential sector only					Residential and commercial sector	
		Total distillate use	Green wood[a]	Green wood[b]	Wood pellets[c]	Pellet raw material if used for drying[d]	Green wood[e]	Pellet raw material if used for drying[d]
		Therms[f]	Tons	Cords	Tons	Cords	Tons	Tons
Aleutians S. Borough	544	523,069	10,256	4,103	4,387	5,331	6,687	8,689
Aleutians W. Borough	1,386	1,240,670	24,327	9,731	10,407	12,644	15,861	20,610
Nome	2,786	3,809,589	74,698	29,879	31,954	38,824	48,703	63,284
North Slope Borough	895	1,751,248	34,338	13,735	14,689	17,847	22,389	29,091
Northwest Arctic Borough	1,825	2,840,027	55,687	22,275	23,822	28,943	36,308	47,178
Wade Hampton Census Area	1,686	2,153,769	42,231	16,892	18,065	21,950	27,534	35,778
Yukon-Koyukuk Census Area	2,135	3,119,137	61,160	24,464	26,163	31,788	39,876	51,814
Total	11,257	15,437,508	302,696	121,078	129,488	157,327	197,358	256,444
Anchorage	3,940	4,126,350	80,909	32,364	34,611	42,053	52,753	68,546
Kenai Peninsula	10,213	11,421,265	223,946	89,579	95,800	116,397	146,013	189,727
Matanuska-Susitna	9,080	9,632,864	188,880	75,552	80,799	98,171	123,150	160,019
Valdez-Cordova	3,226	3,048,145	59,768	23,907	25,567	31,064	38,968	50,635
Total	26,460	28,228,624	553,502	221,401	236,778	287,685	360,884	468,926
Denali	806	1,056,854	20,723	8,289	8,865	10,771	13,511	17,556
Fairbanks North Star	19,922	27,855,719	546,191	218,476	233,650	283,884	356,116	462,732

Table 8. (Continued)

Borough/census area	Adjusted number of housing units	Residential sector only					Residential and commercial sector	
		Total distillate use	Green wood[a]	Green wood[b]	Wood pellets[c]	Pellet raw material if used for drying[d]	Green wood[e]	Pellet raw material if used for drying[d]
		Therms[f]	*Tons*	*Cords*	*Tons*	*Cords*		*Cords*
Southeast Fairbanks	1,757	2,602,632	51,032	20,413	21,830	26,524	33,273	43,234
Total	22,485	31,515,204	617,945	247,178	264,345	321,179	402,900	523,522
Haines	873	754,220	14,789	5,915	6,326	7,686	9,642	12,529
Juneau	7,083	6,074,363	119,105	47,642	50,951	61,905	77,657	100,906
Ketchikan	4,072	2,914,291	57,143	22,857	24,445	29,700	37,257	48,411
Prince Wales–Outer Ketchikan	1,800	1,288,390	25,263	10,105	10,807	13,130	16,471	21,402
Sitka	2,432	1,756,925	34,450	13,780	14,737	17,905	22,461	29,186
Skagway-Hoonah-Angoon	1,515	1,309,491	25,676	10,271	10,984	13,345	16,741	21,753
Wrangell-Petersburg	2,298	1,770,807	34,722	13,889	14,853	18,047	22,639	29,416
Yakutat	347	320,715	6,289	2,515	2,690	3,268	4,100	5,328
Total	20,421	16,189,201	317,435	126,974	135,793	164,988	206,968	268,931
Bethel	4,180	5,338,865	104,684	41,873	44,782	54,410	68,254	88,688
Bristol Bay	633	703,116	13,787	5,515	5,898	7,166	8,989	11,680
Dillingham	1,749	1,995,090	39,119	15,648	16,735	20,332	25,506	33,142
Kodiak Island	3,892	3,449,628	67,640	27,056	28,935	35,156	44,101	57,304
Lake and Peninsula	1,029	1,142,763	22,407	8,963	9,585	11,646	14,609	18,983

Borough/census area	Adjusted number of housing units	Total distillate use	Residential sector only				Residential and commercial sector	
			Green wood[a]	Green wood[b]	Wood pellets[c]	Pellet raw material if used for drying[d]	Green wood[e]	Pellet raw material if used for drying[d]
		Therms[f]	*Tons*	*Cords*	*Tons*	*Cords*		
Total	11,483	12,629,462	247,637	99,055	105,934	128,710	161,459	209,797
Total all	92,106	104,000,000	2,039,216	815,686	872,337	1,059,889	1,329,569	1,727,620

[a] At a combustion efficiency of 60 percent.

[b] At 2.5 tons per cord.

[c] At a moisture content of 6.5 percent and a combustion efficiency of 75 percent.

[d] When part of the raw material is used as an energy source for drying, raw material requirement is increased by up to 25 percent.

[e] Calculated by multiplying the residential amount by 1.63 (multiplier is ratio of commercial oil use in relation to residential oil use; from EIA data (EIA 2008a)).

[f] A therm is a term used to describe 100,000 British thermal units of energy.

Any stimulus that would increase the combustion efficiency of the burning units will reduce the required replacement volumes. British thermal unit equivalent of wood pellets was based on a moisture content of 6.5 percent and a combustion efficiency of 75 percent. This efficiency is slightly less than levels quoted by pellet stove vendors and, again, a higher level of efficiency would result in a decrease in the replacement volume.

Given the above adjustments, it was estimated that the maximum potential annual demand for RWEP to replace distillate fuels for heating in the residential sector of Alaska is about 815,000 cords of green wood or about 872,000 tons of wood pellets (see table 8). If all of the liquid fuels used by the residential and commercial sectors in Alaska were converted to solid wood energy, it is estimated that 1.3 million cords of green wood would be required annually. With respect to wood pellets, a portion of the raw material delivered to the mill is often burned and used to dry the remaining material. Use of part of the raw material as an energy source for drying will increase raw material requirements up to 25 percent. The seventh and ninth columns in table 8 provide an estimate of the volume of material that would be required to produce pellets including material used for drying. The estimated volume of wood that would satisfy the potential demand for wood pellets is about 1.06 million cords of material for the residential sector and about 1.73 million cords of material for both the residential and commercial sectors. This level of RWEP use would represent less than 40 percent of the highest harvest level reported by the Alaska timber industry (Brackley et al. 2009a).

Although table 8 provides potential demand in terms of firewood and pellets, in reality, a portion of the market will be captured by each product. In this report, the demand is based on replaceable fossil fuel Btu. As each product (firewood, pellets, compressed logs, chips, etc.) enters the market, the remaining volume required from other entries is based on the remaining required Btu. In the future, liquid fossil fuel may also be replaced by liquid bio products.

DISCUSSION

Replacement Based on Economics

The conversion among various sources of energy, especially renewable wood energy products, is complex, and price itself may not be the limiting factor.

The conversion among the various sources of energy, especially RWEP, is complex and price itself may not be the limiting factor. Natural gas and electric heating systems are compact and can be used in almost any building. Systems that use fuel oil and other petroleum-base products require tanks, but the space required for tanks is minor. In some areas of extreme cold, these tanks must be located in a partially heated area. On the other hand, renewable wood energy products require considerable storage areas and handling. A year's supply of firewood (assuming up to 8 cords) will require up to 1,100 ft^3 of space (i.e., a building with a floor area of 14 by 14 ft if the wood is piled 6 ft high). Ideally, this storage area should be covered but of sufficient size to allow ventilation to promote drying. In most coastal areas of Alaska, the lowest moisture content resulting from air drying will be 15 to 16 percent green basis. Slightly lower moisture contents may be attained in inland areas such as Fairbanks.

> A year's supply of firewood (assuming up to 8 cords) will require up to 1,100 ft3 of space (i.e., a building with a floor area of 14 by 14 ft if the wood is piled 6 ft high).

One year's supply of wood pellets (6 tons) would require floor area of approximately 100 ft^2 stacked to delivered pallet height. During the heating season, the user must be willing to frequently move fuel from storage to the area where it will be used. Much of the extra work associated with RWEP can be minimized when constructing new homes, designed from the start to utilize these sources of energy, and assuming the evolution of an industry to efficiently deliver the product. Many existing homes, especially in areas of higher population density, were designed for use with fossil fuels or electricity, and this may restrict conversion opportunities.

Table 7 identifies some of the realities when considering replacement of fossil energy sources with RWEP in Alaska. First, given the price for natural gas in the Anchorage-Valley area, there is little price incentive to convert from this source of energy. Natural gas is one of the cleaner burning nonrenewable fossil fuels and in any system to tax carbon, will receive the most favorable treatment of any of the fossil fuels.

The lowest cost for electricity in the populated areas of the state is approximately $0.10/kWh. Electricity produced by hydro, wind, and solar sources is a very clean form of energy. Electricity produced from fossil fuels, however, is a totally different situation. In general, the Btu input of liquid fossil fuel to produce electricity is roughly three times the energy of the resulting electricity. The impact of this is felt in small rural communities where electricity has traditionally been supplied by diesel-powered generators. In such communities, it is not uncommon to find electricity prices of $0.30/kWh or more (AEA 2008). According to EIA (2008c) data, 18 percent of the energy used by the residential sector in Alaska is from electricity. Available census data (U.S. Census Bureau 2006b) indicate that only 10 percent of the Alaska housing units are heated with electricity. Even at the lowest prices for this source of energy, there is an economic incentive to convert to various RWEPs if the housing can be modified to provide fuel storage.

> In general, the Btu input of liquid fossil fuel to produce electricity is roughly three times the energy of the resulting electricity.

Twenty-seven percent of the energy used by the residential sector in Alaska is in the distillate fuels (fuel oil and liquefied petroleum) (see fig. 3). Thirty-eight percent of the housing units are heated with these fuels (see fig. 2). Based on table 7, at a fuel oil price of $2.00/gal the economic incentive to convert to RWEP is minimal, unless the user has a source of standing timber or material that can be converted to firewood using "sweat equity." At a price of $2.00/gal there is little opportunity for a fuel dealer to make a profit from RWEP sales. Although not listed in table 7, the interpolated value of a cord of green wood at $3.00/gal of fuel oil is $270. At this price, the users of "sweat equity" and fuel dealers are both enabled.

At prices above $3.00/gal, there is an obvious economic incentive for homeowners to convert, as the cost saving for RWEP is sufficient to cover the capital costs for converting oil burning equipment to RWEP equipment.

Energy Sources Most Likely To Be Replaced by RWEP

As stated in the "Introduction," replacement of traditional sources of energy for home heating and conversion to RWEP will be a function of a number of factors. The ultimate objectives of this project are to determine initial estimates of the potential volumes that are candidates for replacement and provide an introduction to some of the factors that will impact the rate of replacement and the ultimate amount of replacement.

In addition to identification of factors, there are many questions relative to how the factors will interact to promulgate change. It is certain that conversion to RWEP will take place over a period of years. The Alaska Housing Manual (AHFC 2000) reported that the most common heating system used in the Anchorage-Valley region were centralized, gas fired, hot air systems. In other areas of the state, centralized oil fired hydronic systems are the norm. In new home construction or in upgrades of heating systems, it is possible to integrate fossil fuels and RWEP into a common air or water distribution system. Burners that incorporate oil, solid wood, or pellets in one integrated unit or as two independent units are available. There would be more incentive to install multifuel systems where high-priced sources of energy are used. In the long term, transparent (i.e., requiring relatively minor changes to burner system and fuel delivery method) conversion to liquid RWEP is an option for homeowners with fuel-oil based systems. It is more likely, however, that most of the initial conversion to RWEP will involve increased use of space heaters and fireplace inserts, thus reducing the dependence on the high-cost energy alternatives.

The ultimate objectives of this project are to determine initial estimates of the potential volumes that are candidates for replacement and provide an introduction to some of the factors that will impact the rate of replacement and the ultimate amount of replacement.

Sources of Renewable Wood Energy Material

Some individuals may assume that conversion and utilization of RWEP in accordance with the previously stated numbers will result in a cord-for-cord increase in the harvest levels from the forests of the state. For many reasons, however, this is not the case. A complete analysis and statement of the reasons why the assumption is incorrect is beyond the scope of this paper, but a few comments are appropriate.

First, the U.S. Department of Agriculture, Forest Service, Forest Inventory and Analysis (FIA) Program is charged by Congress with maintaining an inventory of timber volumes in the forest (public and private) of the United States. Trees defined as timber include stems that can currently be harvested into saw logs, those that have the potential to grow into logs at some future date, and trees that owing to poor form (i.e., they are crooked) or rot, cannot produce a saw log. By definition, growth takes place only on the growing-stock trees—those that currently are suitable for producing saw logs or have the potential to grow into saw logs. Given these definitions, if a tree that does not include a saw log or have the potential to grow into a saw log is removed from the woods and used, the removal has no impact on growth as defined by the FIA. In the past, timber volumes have been defined in terms of merchantability standards. The standards of the past excluded material in tops and limbs left in the woods from the inventoried timber volumes. Tops, limbs, rough and rotten trees, and trees below standards of merchantability, however, can all be harvested for fiber and converted into energy products.

Also, residual products such as slabs, edgings, or sawdust from saw logs can be converted into energy products and the use of residuals does not result in any increase in the number of trees cut and harvested. The size of a residual-based industry, however, is somewhat limited by the capacity of the mills to process the solid wood product, such as lumber.

Many homeowners produce firewood from trees that do not grow in the forest or from dead and low-grade trees that are not considered as forest growing stock. Homeowners can also go into areas that have been logged and cut firewood from branches and tops that are left in the woods as slash. Use of these materials does not have any impact on growth.

> If high-quality trees of saw-log quality or trees that have the potential to grow into sawlog-quality trees are harvested and used for energy products, there will be a reduction in growing stock material.

These are a few of the situations that need to be taken into account to determine the impact of energy use on the forest. As a rule of thumb, if high-quality trees of saw-log quality or trees that have the potential to grow into saw-log-quality trees are harvested and used for energy products, there will be a reduction in growing stock material. If, however, the material is from any other source (e.g., land not considered part of the forest, trees not of growing-stock quality, or parts of trees already harvested), it is in fact considered an increase in utilization of harvested material that has no impact as far as timber sustainability is concerned. Such changes in utilization may, however, have an impact on the sustainability of forest ecosystems when the concepts of biodiversity and related value of course woody debris are considered. A complete review of these issues is beyond the scope of this project.

CONCLUSION

The level and satisfaction of future demand for RWEP in Alaska really has little to do with the existence of a biomass resource. It is limited, however, by the size of the existing forest products industry, the industry's capacity to economically harvest wood, and society's willingness to convert. In addition, conversion to RWEP will be a function of a national energy policy and price for alternatives. Lacking a national energy policy, a return to high market prices for oil will stimulate production of energy products and result in conversion. It is also possible that conversion will in part promote the development of a more vibrant forest products economy in Alaska.

Total conversion of oil and other liquid fuels used by the Alaska residential and commercial sectors to solid RWEPs would require in excess of 1.3 million cord equivalents of material annually. Although that volume may appear great to many people, in reality it represents the amount of wood required to supply raw material to one large pulp mill.

The economic incentive to convert to solid wood fuel exists at any heating oil price in excess of $3.00/gal. At this level, fixed costs are recovered in relatively short periods (5 years or less). A national energy policy may impact conversion by placing a tax on fossil fuels or providing tax credits to help cover the costs of converting to systems that use RWEPs (biomass).

> Total conversion of oil and other liquid fuels used by the Alaska residential and commercial sectors to solid renewable wood energy products would require in excess of 1.3 million cord equivalents of material annually.

ACKNOWLEDGMENT

This report is based upon work supported by the University of Alaska and Cooperative State Research, Education and Extension Service, U.S. Department of Agriculture, under Agreement No. 2006-34158-17722. Any opinions, findings, conclusions, or recommendations expressed in this publication are those of the author(s) and do not necessarily reflect the view of the U.S. Department of Agriculture. Funds to support the Fairbanks Economic Development Corporation, Wood Pellet Market Survey activity were provided in part by USDA Forest Service Joint Venture Agreement PNW-07-JV-11261935-042 between the U.S. Forest Service, Pacific Northwest Forest

Research Station, Alaska Wood Utilization Center, and the Fairbanks Economic Development Corporation.

METRIC EQUIVALENTS

When you know:	Multiply by:	To find:
British thermal units (Btu)	1,050	Joules
Inches (in)	2.54	Centimeters
Feet (ft)Cubic feet (ft3)	.305 .0283	Meters Cubic meters
Miles (mi)	1.609	Kilometers
Gallons (gal)	3.78	Liters
Thousand board feet, log scale (mbf)	4.5	Cubic meters, logs
Tons (ton)	907	Kilograms
Pounds (lb)	454	Grams
Degrees Fahrenheit	.56(°F − 32)	Degrees Celsius
British thermal units	3,412.14	Kilowatt-hour (kWh)

COMMON AND SCIENTIFIC NAMES

Common name	Scientific name
Alaska yellow-cedar	*Chamaecyparis nootkatensis (D. Don) Spach*
Black cottonwood	*Populus balsamifera L. ssp. trichocarpa* (Torr. & A. Gray ex Hook.) Brayshaw
Paper birch	*Betula papyrifera Marsh.*
Quaking aspen	*Populus tremuloides Michx.*
Red alder	*Alnus rubra Bong.*
Sitka spruce	*Picea sitchensis (Bong.) Carriere*
Western hemlock	*Tsuga heterophylla (Raf.) Sarg.*
Western redcedar	*Thuja plicata Donn ex D. Don*
White spruce	*Picea glauca (Moench) Voss*

REFERENCES

Alaska Climatic Research Center [ACRC]. 2009. Mean annual heating degree days for selected bases (1971–2000). http://climate.gi.alaska.edu/Climate/ Normals/HDD.html. (March 17, 2009).

Alaska Energy Authority [AEA]. 2008. Statistical report of the power cost equalization program, fiscal year 2007. 19th ed. February 2008. Juneau, AK: State of Alaska. 22 p.

Alaska Housing Finance Corporation [AHFC]. 2000. Alaska housing manual. 4th ed. Anchorage, AK: Research and Rural Development Division. 119 p. http://www.ahfc.state.ak.us/reference/housing_manual.cfm. (April 6, 2009).

Bowyer, J.L.; Shmulsky, R.; Haygreen, J.G. 2003. Forest products and wood science—an introduction. 4th ed. Ames, IA: Iowa State University Press. 554 p.

Brackley, A.M.; Haynes, R.W.; Alexander, S.J. 2009a. Timber harvests in Alaska: 1910–2006. Res. Note PNW-RN-560. Portland, OR: U.S. Department of Agriculture, Forest Service, Pacific Northwest Research Station. 24 p.

Brackley, A.M.; Nicholls, D.L.; Hannan, M. 2009b. An evaluation of the grades and value of red alder lumber in southeast Alaska. Gen. Tech. Rep. PNWGTR-774. Portland, OR: U.S. Department of Agriculture, Forest Service, Pacific Northwest Research Station. 27 p.

Briggs, D. 1994. Forest products measurements and conversion factors: with special emphasis on the U.S. Pacific Northwest. IFR Contribution No. 75. Seattle, WA: University of Washington, College of Forest Resources. 168 p.

Bruce, D.; Schumacher, F.X. 1950. Forest mensuration. 3rd ed. New York: McGraw-Hill Book Company, Inc. 483 p.

Dunster, J.; Dunster, K. 1996. Dictionary of natural resource management. Vancouver, BC: UBC Press. 363 p.

Electronic Code of Federal Regulations [ECFR]. 2009. Title 40: protection of environment. Part 60—standards of performance for new stationary sources.http://ecfr.gpoaccess.gov/cgi/t/text/text-idx?c=ecfr;sid=c3e6eb 575293d0041e7a768e297b67d9;rgn=div6;view=text;node=40%3A6.0.1.1. 1.65;idno=40;cc=ecfr. (June 16, 2009).

Energy Information Administration [EIA]. 2008a. Alaska—state energy data system (SEDS): production, 1960–2006. http://www.eia.doe.gov/emeu /states/state.html?q_state_a=ak&q_state= ALASKA. (September 3, 2008).

On file with: Allen Brackley, Pacific Northwest Research Station, Alaska Wood Utilization Research and Development Center, 204 Siginaka Way, Sitka, AK 99835.

Energy Information Administration [EIA]. 2008b. Consumption, price, and expenditure estimates—state energy data system. http://www.eia.doe.gov /emeu/ states/_seds.html. (November 2008).

Energy Information Administration [EIA]. 2008c. State energy profiles— Alaska. http://tonto.eia.doe.gov/state/state_energy_profiles.cfm?sid=AK. (September 3, 2008).

Energy Information Administration [EIA]. 2008d. Technical notes and documentation—state energy data system (SEDS). http://www.eia.doe. gov/emeu/states/_seds_tech_notes.html. (September 3, 2008).

Eshleman, C. 2008. Fairbanks Borough mayor pushes for wood stove trade-in plan. Fairbanks Daily News-Miner. December 8. http://www.newsminer. com/ news/2008/dec/08/fairbanks-borough-mayor-jim-whitaker-pushes-wood-s/. (April 8, 2009).

Evans, D.S., ed. 2000. Terms of the trade. 4th ed. Eugene, OR: Random Length Publications, Inc. 425 p.

Husch, B.; Miller, C.I.; Beers, T.W. 1982. Forest mensuration. 3rd ed. New York: John Wiley & Sons. 402 p.

Ince, P.J. 1979. How to estimate the recoverable heat energy in wood and bark fuels. Gen. Tech. Rep. FPL-29. Madison, WI: U.S. Department of Agriculture, Forest Service, Forest Products Laboratory. 6 p.

Kollmann, F.F.P.; Cote, W.A. 1968. Principles of wood science and technology. Part 1 Solid wood. New York: Springer-Verlag. 592 p.

National Appliance Energy Conservation Act of 1987 [NAECA]. 42 U.S.C. Pub. L. 100-12; 101 Stat. 103.

Nicholls, D.; Brackley, A. 2009. House log drying rates in southeast Alaska for covered and uncovered softwood logs. Gen. Tech. Rep. PNW-GTR-782. Portland, OR: U.S. Department of Agriculture, Forest Service, Pacific Northwest Research Station. 18 p.

Nicholls, D.; Monserud, R.A.; Dykstra, D.P. 2009. International bioenergy synthesis—lessons learned and opportunities for the Western United States. Forest Ecology and Management. 257(8): 1647–1655.

Oregon Department of Forestry [ODF]. 2009. Log price information. http://www.oregon.gov/ODF/STATE_FORESTS/TIMBER_SALES/logpa ge. shtml. (April 9, 2009).

Patton-Mallory, M., ed. 2008. Woody biomass utilization strategy. FS-899. Washington, DC: U.S. Department of Agriculture, Forest Service. 17 p.

Robb, S. 2007. Survey to assess community support for pellet fuel—a report prepared for the Fairbanks Economic Development Corporation. Anchorage, AK: Information Insights, Inc. 36 p.

U.S. Census Bureau. 2000. American FactFinder, DP-4. profile of selected housing characteristics: 2000. http://factfinder.census.gov/servlet/ QTTable?_bm=n&_lang=en&qr_name=DEC_2000_SF3_U_DP4&ds_na me=DEC_2000_ SF3_U&geo_id=04000US02. (July 1, 2008).

U.S. Census Bureau. 2004. Housing data between the censuses: the American housing survey. Census Rep. AHS/R/04-2. Washington, DC. 23 p.

U.S. Census Bureau. 2005a. American housing survey. http://www.census. gov/hhes/www/housing/ahs/ahs.html. (July 1, 2008).

U.S. Census Bureau. 2005b. Cartographic boundary files. http://www.census. gov/ geo/www/cob/tr2000.html. (June 16, 2009).

U.S. Census Bureau. 2006a. Design and methodology: American community survey. Tech. Pap. 67. Washington, DC: U.S. Government Printing Office. 420 p.

U.S. Census Bureau. 2006b. State and county QuickFacts. http://quickfacts. census. gov/qfd/states/00000.html. (September 3, 2008).

U.S. Department of Agriculture, Forest Service [USDA FS]. 1999. Wood handbook—wood as an engineering material. Gen. Tech. Rep. FPL-GTR-113. Madison, WI: Forest Products Laboratory. 463 p.

Wilson, P.L.; Funck, J.W.; Avery, R.B. 1987. Fuelwood characteristics of northwestern conifers and hardwoods. Res. Bul. 60. Corvallis, OR: Oregon State University, College of Forestry, Forest Research Lab. 42 p.

INDEX

D

E

S

T